The City as Fulcrum of Global Sustainability

The City as Fulcrum of Global Sustainability

by

Ernest J. Yanarella and Richard S. Levine

ANTHEM PRESS
LONDON · NEW YORK · DELHI

Anthem Press
An imprint of Wimbledon Publishing Company
www.anthempress.com

This edition first published in UK and USA 2011
by ANTHEM PRESS
75-76 Blackfriars Road, London SE1 8HA, UK
or PO Box 9779, London SW19 7ZG, UK
and
244 Madison Ave. #116, New York, NY 10016, USA

British Library Cataloguing-in-Publication Data
A catalogue record for this book is available from the British Library.

Library of Congress Cataloging-in-Publication Data
The city as fulcrum of global sustainability / by Ernest J. Yanarella and Richard S. Levine.
p. cm.
Includes bibliographical references and index.
ISBN-13: 978-0-85728-772-4 (hardback : alk. paper)
ISBN-10: 0-85728-772-9 (hardback : alk. paper)
1. Urban ecology (Sociology) 2. Sustainable development.
3. Cities and towns–Case studies. 4. Sustainability. I. Levine, Richard S. II. Title.
HT241.Y37 2011
307.76–dc23
2011028420

ISBN-13: 978 0 85728 772 4 (Hbk)
ISBN-10: 0 85728 772 9 (Hbk)

This title is also available as an eBook.

TABLE OF CONTENTS

Closing Thoughts

Appendix

PREFACE AND ACKNOWLEDGMENTS

This book originated around 1990, when the two authors crossed paths on the main campus of the University of Kentucky and struck up a conversation about solar energy advances and the idea of sustainable cities. In spite of the gulf separating our disciplines (political science and architecture), we found an enormous overlap in interests in the solar movement and sustainability. Out of that stimulating discussion came a mutual commitment to engage in a series of periodic meetings guided by an evolving list of readings that might help clarify our common grounding in going beyond solar energy and conservation to a vision of sustainable cities emanating from the model of the polis (Yanarella) and the ideal of the medieval Italian hilltown (Levine). This reading and discussion process led to the penning of a sustainable city manifesto and to involvement in contributing to an international charter through Levine's participation in the Aalborg conference in Denmark as keynote speaker. From there, the authors achieved modest university funding to support a multidisciplinary seminar and lecture series, which culminated in the establishment of the Center for Sustainable Cities as a joint research and policy center under the aegis of the university's Colleges of Arts and Sciences and Architecture.

The substantive chapters in this book are just part of the theoretical and design work generated from our individual and collaborative efforts over more than twenty-five years. While our programs and projects have gained only modest notice in the United States, their influence has been more substantial in Europe (particularly Italy and Austria, but also Germany), the Middle East and China. The development of a sister institution – Oikodrom: the Vienna Institute for Urban Sustainability – in Austria in 1994 under the able direction of Dr Heidi Dumreicher enabled us to get a foothold on a continent, where sustainability work has long been ahead of Anglo-American research. It also provided us with a string of collaborative research opportunities in the European Union that has advanced an urban sustainability agenda that has not been possible in the United States. One of our aspirations is that an Anglo-American audience reading this book will discover substance and value in the theory, design and method of sustainable city-regions that we have painstakingly tried to articulate over two and a half decades.

We have many people to thank for their witting and unwitting contributions to the efforts that have culminated in this book.

I (Yanarella) wish to express my appreciation of the generation of graduate students at the University of Kentucky whose doctoral work carried the writing and design studies into new and fertile territory. These include: Robert W. Lancaster, Hugh Bartling, Christopher Rice, Donald Ferris and D. Gene Pace. I am proud of their involvement in PS 776 (Sustainable Urban Design) and subsequent coursework with me, and I am gratified that my intellectual-political commitments and theoretical vision have contributed to their further intellectual and social development and career trajectories. It is my hope that they will continue to plumb the writings of my mentors – Lewis Mumford, Patrick Geddes, Ernst Bloch and Antonio Gramsci – as they move into the mid-career stage.

I am grateful for support from a variety of funding sources, especially the Canadian Studies Program, which permitted frequent travel to Canadian sites to study their urban sustainability programs, as well as the Kentucky Environmental Education Council, which recently supported valuable curricular and public education activities at the University of Kentucky and around the Bluegrass. To the University of Kentucky College of Arts and Sciences and the Dean's Office, I want to express my gratitude for being selected as the Arts and Sciences Distinguished Professor for 2009–10. This coveted award gave me a paid semester off from teaching to finish this book and, in addition, subsidized some of the costs of its publication.

I also wish to thank Lori Tripoli for the opportunity to serve on the editorial board and contribute to *Sustainability: The Journal of Record*. I would also like to thank Matthew McCauley and Victor Shanker of the firm GeoSim, who have allowed me to develop high-level contacts in the context of joining them in presentations in Canada and the United States promoting their powerful 3D visualization software and services. While my co-author (Levine) believes Building Information Modeling software is the way to achieve our dream of a Sustainability Engine, I would put my money on GeoSim as the best evolving software to bring the Engine to fruition. For intellectual comradeship that has spanned decades, I owe a deep debt to Herbert G. Reid, who always seems to manage to think further into the theoretical horizon on important and compelling issues of democratic theory, global sustainability and social justice and to share so unselfishly his many profound theoretical and political insights over a lifetime with me and so many others. To my sustainability comrade in arms and coauthor, Richard S. Levine, I can never fully express what a deep and enriching experience it has been walking with him on this intellectual journey and how blessed I feel for his seriousness and generosity in so much we have tried to do to advance the cause of urban sustainability. Finally, and not least of all, I thank my wife and soulmate, Elizabeth Walker Yanarella, for her love and support, as well as her many contributions in music and carpentry to sustaining our marriage, our family and our household economy.

I (Levine) wish first to extend my gratitude to the many generations of talented architecture students who accompanied me on this journey only to discover in my design studios either a compelling refuge from the prevailing fashion of the day or a great disaffection with the same. Many of the former have become lifelong friends and colleagues. Principal among them are Bob Koester and Bill Murrell, early students and collaborators who have each made significant contributions; Taqi Radmard, a frequent and brilliant partner over the years; and my current associates, Michael Hughes and Casey Ryan Mather, who have great careers ahead of them.

In many ways, I owe the most to three powerful women. My late wife Anne Frye, a fine artist, provided the grounding to see the world as it should be and the terror to seek a realistic simulacrum of that impossible place. To my dear daughter Laura Alex Frye-Levine, who is emerging as an environmental force in her own right, I owe the gratitude of her loyalty to the person and the cause, and the forthrightness of her ideas and critique in spite of – or perhaps because of – my many shortcomings. I would like to thank my long-time colleague and dearest friend Heidi Dumreicher for the amazing things we have sometimes accomplished through her tolerance or intolerance of our mutual compatibilities or incompatibilities.

The rock of my journey to this new world of sustainability has been my long-time colleague Ernie Yanarella, who has always been there with both his ideas and critique, as well as his mastery and understanding of the intellectual basis of our common work.

In the end, I feel compelled to acknowledge an enduring debt to my colleagues and administrations at the University of Kentucky College of Architecture, whose diligent pursuit of prevailing paradigms over many decades has given this slow thinker the time and the solitude to imagine and rehearse an alternative world as it could be, should be and finally must be.

We would be negligent if we did not thank Janka Romero and Tej P. S. Sood for their editorial guiding hands at Anthem Press. Their able stewardship of the book manuscript through the review and publication process, enormous patience in dealing with delays and good cheer through the process have sustained us throughout our labors. Our copyeditor, Florence Huffman, has deployed her perfectionist talents in translating our often arcane academic writing and technical language into fluid prose and a consistent style, for which we are deeply appreciative. Thanks too to Erik Davidshofer for taking on much of the typing and assembling the bibliography, without which this work would otherwise be the poorer.

Finally, we express our thanks to the following journals and their editors for giving us permission to incorporate all or part of previously published journal articles into this book:

Chapter 1: Yanarella, Ernest J. and Richard S. Levine. "Does Sustainable Development Lead to Sustainability?" *Futures* 24 (October 1992): 759–74. Elsevier Ltd.

Chapter 2: Yanarella, Ernest J. and Richard S. Levine. "The Sustainable Cities Manifesto: Pre-Text, Text, and Post-Text." *Built Environment* 18 (December 1992): 301–13. Alexandrine Press.

Chapter 3: Yanarella, Ernest J., Richard S. Levine and Robert Lancaster. "Green vs. Sustainability: From Semantics to Enlightenment." *Sustainability: The Journal of Record* 2 (October 2009): 296–302. Mary Ann Liebert, Inc.

Chapter 4: Yanarella, Ernest J. and Richard S. Levine. "Don't Pick the Low-Hanging Fruit! Counterintuitive Policy Advice for Achieving Sustainability." *Sustainability: The Journal of Record* 1 (August 2008): 256–61. Mary Ann Liebert, Inc.

Chapter 6: Yanarella, Ernest J. and Richard S. Levine. "The Space of Flows, the Rules of Play, and Sustainable Urban Design: The Sustainability Game as a Tool of Critical Pedagogy in Higher Education." *International Journal of Sustainability in Higher Education* 1 (2000): 48–66. Emerald Publishing Group, Ltd.

Chapter 9: Yanarella, Ernest J. "Sustainability Comes to the Canadian Prairie: Lessons and Caveats from the Town of Okotoks, Alberta." *Terrain.org: A Journal of the Natural & Built Environments* 18 (Spring/Summer 2006): 191–208. Simmons Buntin, Terrain.org owner and editor.

Chapter 11: Levine, Richard S., Michael T. Hughes, C. Ryan Mather and Ernest J. Yanarella. "Generating Sustainable Towns from Chinese Villages: A System Modeling Approach." *Journal of Environmental Management* 87(April 2008): 305–16. Elsevier Ltd.

Ernest J. Yanarella and Richard S. Levine
Lexington, Kentucky
April 2011

LIST OF TABLES AND FIGURES

TABLES

FIGURES

INTRODUCTION

An Unsustainable World in Crisis

Between 2001 and 2009, five signal events shaped the contemporary local-national-regional and global scene in the first decade of the new millennium in ways that will reverberate throughout the world for years, perhaps decades, to come. These events were: the war on terrorism precipitated by the events of September 11, 2001; the shift in global population from the rural countryside to towns and cities heralded by the United Nations (UN); the Intergovernmental Panel on Climate Change's Assessment Report declaring that the evidence for global warming was unequivocal; the global economic crisis triggered by the financial meltdown of major banks and insurance firms beginning in July and cresting in early September 2008; and the prospect of reaching peak oil production between 2008 and 2010, coupled with the specter of a multitude of other resource scarcities.

Since 9/11, the global war on terrorism has cast a long shadow over the shape of world politics and American foreign and domestic politics. It has threatened – and continues to threaten – to marginalize the national environmental agenda, spawned by the movement politics of the sixties and informed by the legacy of the progressive conservationism and preservationist schools of the early twentieth century. Meanwhile, the Intergovernmental Panel on Climate Change (IPCC) released its fourth major report on global warming in early 2007, which stated that warming of the climate system is unequivocal and attributing the rise in average global temperatures with 90 percent likelihood to anthropogenic (i.e., human and social) sources. In its report, the IPCC warned that such warming and sea-level rise would likely continue for centuries, depending upon national and globally coordinated mitigation and adaptation policy action (IPCC, 2007).

While greeted with media fanfare and noted by national and international policymakers and scholarly musings, the majority shift in world population to cities somewhere between 2007 and 2008 was generally met with relative indifference by citizens around the world (UN-HABITAT, 2006: 6). This historic landmark has led to a dramatic turnaround by the World Bank, which has recently shown renewed policy concern and focus upon the character and impact of global urbanization processes. Likewise, various UN agencies, among them the UN Population Fund, have issued projections and reports that see these urbanization trends both in terms of promise and challenge, where these changes serve as harbingers of new opportunities for urban-driven economic development; as improvements in the quality of life of inhabitants of the less-developed world; and as threats to global sustainability in the face of continued (but slowing) growth among the world's 20 megacities. These urbanization trends also provide for more widely distributed pressure for economic growth and satisfaction of basic human needs in cities and towns in Africa and Asia, where the bulk of heightened population increases will be centered.

The global economic slowdown, which followed the United States' financial tailspin in the late summer and early fall of 2008, has gripped national economies around the world, dried up large-scale sources of capital for promoting major sustainability projects, and hamstrung national political leaders and legislators in funding new green initiatives to advance sustainability goals (Foster and Magdoff, 2009; Krugman, 2009; and Stiglitz, 2010). This triple condition has created in the United States what Krueger and Agyeman (2005) have called a "sustainability schizophrenia," where the dawning of the Urban Millennium, the War on Terror and the global financial meltdown have shifted national and international policy attention away from liveability and sustainability issues toward the pressing problems of putting the world's banking and financial system back in order, igniting sustainable growth rates and putting the newly unemployed back to work. Meanwhile, efforts at the local level in cities and towns in the United States, Canada and elsewhere across the globe, in part catalyzed by the looming problem of climate change, continue full-speed, as so-called "green initiatives" proliferate throughout business management and consumer production and extend to virtually all other realms of everyday life.

Finally, both independent and oil company analysts have begun to take seriously estimates of world peak oil production projected by M. King Hubbert's methodology, models and calculations that had accurately forecasted the topping off of oil production in the United States between 1965 and 1970 (Hubbert, 1962; Deffeyes, 2005). While sagging oil consumption, triggered by the significant global economic slowdown in late 2008, may marginally shift predictions of world peak oil production by a few years, the dawning of a "long emergency" in energy supplies (Kunstler, 2005) and the scramble for remaining finite natural resources to fuel the industrial world and an economically growing China and India seem inevitable. These factors also promise to pose daunting problems for economic stability and national security (Kalicki and Goldwyn, 2005). While history demonstrates that in such contexts war often becomes the ultimate arbiter, a more hopeful scenario for the future lies in the sustainability alternative, where nations hasten to

Figure 0.1. National, regional and global peak oil production (billions of barrels/year)

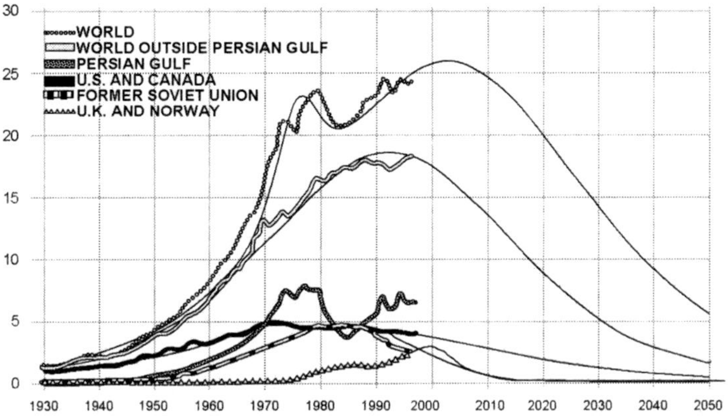

Source: Campbell and Laherrere, 1998: 81

steer individual courses and collective actions toward a postpetroleum world involving new and innovative practices and processes grounded in renewable resources and conservation.

Sustainability in an Urbanizing World

We have written this work in an effort to unite concerns about the continuing urbanization of the community of nations and the implications of the shift of human populations to the world's cities for achieving a higher quality of life and standard of living, greater liveability for inhabitants of cities and true urban sustainability as the means for rebalancing and reconciling humankind's ineluctable embeddedness in its earthly host. As a contribution to Anthem Press's "Environmental Studies" series, we fundamentally believe it serves an important role in extending the limits of twenty-first century environmentalism by fully incorporating the themes of urbanization and sustainability, two relative newcomers to the environmental studies field, into its fold. It seeks to do so in a style offering rigorous critique, intellectual advocacy and practical example.

The winds of war spawned by the terrorist acts of 9/11 and the economic fallout of the global financial crisis have crowded out sobering events signaling a continuing environmental crisis of global proportions. Meanwhile, the efforts of grassroots organizations at the local and regional levels, the flowering of new social movements and an array of green programs and activities at the end of the first decade of the new millennium all point to the continuing vibrancy of the multicultural conversation and global debate over strategies of ecological and social sustainability. Begun in 1987 by the Brundtland Commission report on Earth's common future and further promoted by the Rio Earth Summit (1992), this worldwide dialogue refuses to be silenced in the face of both the mobilized state of the war on terrorism and the globalization of the financial cataclysm. Indeed, with every scientific report focusing on global warming and the deterioration of the ecosystem, the sustainability debate threatens to break out of its media confinement and burst forth with renewed energy and public focus.

The many different versions of sustainability articulated in scholarly and popular writings, town hall forums and international conferences have both enriched and hobbled the sustainability movement from the grassroots to the global level in a manner analogous to the times featured in Charles Dickens' *A Tale of Two Cities* where:

> It was the best of times, it was the worst of times, it was the age of wisdom, it was the age of foolishness, it was the epoch of belief, it was the epoch of incredulity, it was the season of Light, it was the season of Darkness, it was the spring of hope, it was the winter of despair… (Dickens, 1859: 13)

In the context of the stark realities and contradictions that populate our twenty-first century world, the dilemmas and seeming paradoxes of a

Table 0.1. Summary of global trends in terrorism and demographics

Trends in Terrorism	
#1	The war on terror has not reduced terror
#2	The threat of Islamist terror has changed
#3	Al Qaeda increasingly sees it mission today as fueling radicalization worldwide
#4	Terrorism is globalizing
#5	We understand more about Islamic radicalization than ever before
Demographic Trends	
#1	A bulge of young people in Islamic countries
#2	Large populations hinder counterterrorism
#3	There will be many more young Muslims living in weak or failed states
#4	Islamic immigration into Europe will grow

planet – marked by extraordinary wealth and abject poverty; incredible scientific progress and woeful ignorance; swollen military expenditures and declining social-service and foreign-aid budgets; countless examples of human kindness and generosity; and innumerable instances of inhuman cruelty and profound ethnic, religious, and racial hatred toward the other – present themselves. Reflecting on these tensions and contradictions, we ask if the gathering energy and intelligence being directed to solving the world's most portentous and knotty problems of living together, sharing the Earth's resources, building a world of elegant frugality and durable peace and addressing the urbanization of the globe's still mounting population can be advanced by embracing sustainability as the answer. Without making the urban sustainability enterprise tolerate more weight than it can bear as a solution, we argue that a world of sustainable cities and city-regions would do much to take pressure off the ecological burdens of the built environment and the wider ecosystem manifested in continuing acid rain, incipient climate change, diminishing potable water supplies, eroding top soil, peak oil and other rising ecological scarcities. We claim that despite the many loud and seemingly contrary voices comprising the world's sustainability conversation, such a world of sustainable cities and city-regions is not a utopian dream, but a political imperative that can be mightily furthered by a type of strong sustainability strategy that we outline in this book. After limning some of its key elements, this introductory chapter closes with a preview of the individual chapters comprising the book and the central argument holding them together.

Sadly, the present state of popular images and understandings of the dimensions of the sustainability debate remain caught in a web of myths and misconceptions. That is, the popular debate in the news media and mass culture tends to be "hung up on" the apparent contradiction between the dynamics of national economic growth and a global-political economy, on the one hand, and the needs and demands for ecosystem well-being, environmental protection and a livable and sustainably built environment, on the other. As a result, the sustainability debate as presented in the mass media often flounders on a set of distorted public understandings and manufactured half-truths about the implications of sustainability for societal economic well-being and individual and collective quality of life. It also leads to bogus policy resolutions that seek either to balance pressing demands for economic growth against costly environmental necessities or to be overcome by escalating green nostrums that amount to "picking the low-hanging fruit."

By contrast, this volume seeks to peel away semantic confusion and ideological fog on this vital public issue and replace them with a lucid understanding of the competing schools of thought populating the debate over sustainability and the way beyond this loud and raucous dispute. The coauthors are partisans in the sustainability debate who have contributed new perspectives and tools in the pursuit of a world and a built environment in harmony with nature. For this reason, the volume will ultimately defend a point of view and outlook on the entwining of urbanization and sustainability.

Sustainability as an "Essentially Contested Concept"

Since its rise to prominence in community initiatives and public policy formulations, sustainability has gained its popularity and allure for a number of reasons, not least of which is its vagueness and plasticity. The Brundtland Commission's definition is a perfect example of this point. Indeed, so many people have come to embrace the idea of sustainability because of its openness and multiple understandings. In a word, sustainability is an "essentially contested concept."

The principle of essential contestability was first advanced by W. B. Gallie (1956). For Gallie, a concept may be suspected of being essentially contested when its meaning leads to unending disputes without apparent resolution. William Connolly suggests that a *prima facie* case for the essentially contested character of a concept can be made when: (1) it appraises a state of affairs and is seen as a "valued achievement"; and (2) this practice is "internally complex," having a number of facets or dimensions, and its defining qualities are "relatively open," leading many people to interpret its meaning in new contexts in different ways (Connolly, 1983: 10).

"Essentially contested concepts," however, should not lead to the conclusion that varying interpretations of these concepts are simply arbitrary positions set forth without rational basis. Instead, contestants disputing such concepts like power, freedom, justice – and, yes – sustainability are (or should) find themselves led to reevaluate their own positions in light of the arguments and evidence, which are brought to bear by other parties, thus contributing "to the intellectual development of all protagonists" (Connolly, 1983: 1). That is, the rational outcome of honest and serious debate over such categories should be greater understanding of one's preferred definition and a wider appreciation of the complex and nuanced nature of the concept at hand.

As Gallie and Connolly well observe, essentially contested concepts are better understood as *appraisive* in character in that they seek both to *describe and give value to* the referent they examine. So if we say that a building or a city or a region or society is unsustainable, we are characterizing it "from the vantage point of certain interests, purposes, or standards" (Connolly, 1983: 28); i.e., we are characterizing it from the standpoint of explicitly (or more often tacitly) accepted standards of sustainability, ecological well-being or the like.

This approach then clarifies a fundamental facet of these contests; specifically, the way in which these disputes are not merely about the concept of sustainability but are constitutive of sustainability itself. To convince others to embrace my interpretation of an appraisive concept (e.g., sustainability) is "to implicate them in judgments to which I am committed and to encourage [such] activity congruent with these commitments" (Connolly, 1983: 30). Indeed, part of the intent of this book is to try to convince supporters of sustainability to see the deeper meaning and substance of sustainability in a manner convergent with the strong reading of sustainability that we have provided in our definition.

Conceptual disputes are often contested because disputants feel others only partly understand the concept's meaning, but simply do not quite get the rest. If we had no handle on the meaning of sustainability beyond our individual paradigms, we would become hopelessly caught up in nihilism or solipsism; if we completely grasped its meaning (without remainder or opaqueness), we would not need to strive to incorporate genuine insights about sustainability from those opened up in dialogue with others.

Let us continue with a line of inquiry concerning the import of politics in disputes over essentially contested concepts like sustainability. We might ask: why are conceptual debates often so hotly disputed? Operating in a Galilean spirit, William Connolly offers us a better way of answering these queries. For him, the essentially contested nature of a spectrum of concepts points to the ineluctable political dimension of such battles. In his words:

> When groups range themselves around essentially contested concepts, politics is the mode in which the contest is normally expressed. Politics involves the clash that emerges *when appraisive concepts are shared widely but imperfectly*, when mutual understanding and interpretation is possible but in a partial and limited way, when reasoned argument and coercive pressure commingle precariously in the endless process of defining and resolving issues [emphasis added]. (Connolly, 1983: 40)

So, there is a politics of essentially contested concepts and that politics is resolved through the deployment of argument and power in the pitch of the linguistic and conceptual confrontation.

Sustainability: New Concepts, New Theory, New Operational Definition

If sustainability is an essentially contested concept, how should one retrieve it from the morass of ever-expanding and increasingly vague and mushy

Figure 0.2. Sustainability circles and tripod

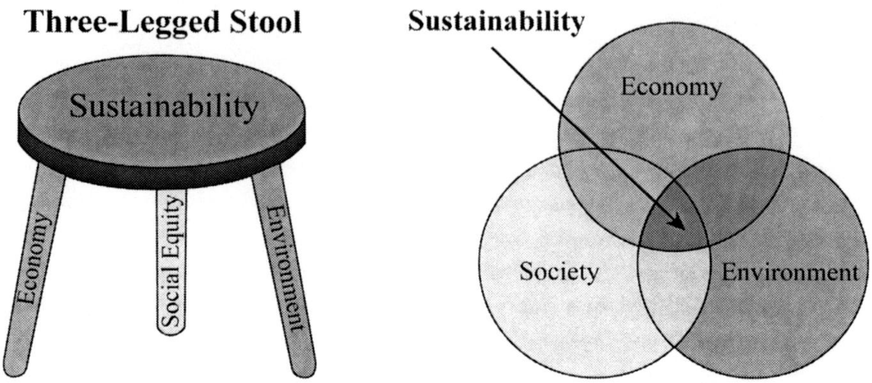

Source: rendered by Tom Dolan at the University of Kentucky Academic Technology Group

understandings of the term? In the face of these rapidly proliferating meanings, both explicit and implicit, how can we ground sustainability in a solidly substantive and persuasive interpretation that allows bona fide scientific and social scientific research to be conducted on it?

Any definition of sustainability or sustainable development must begin with (or take as its point of departure) two notions. The first is the sustainability tripod or triple bottom-line. By essence or convention, sustainability has been identified with three legs or overlapping circles that stand for environmental health, economic well-being and social justice. This metaphoric way of representing sustainability, we argue, is perhaps the first word, though certainly not the last word, in grasping key features of sustainability or sustainable development. Its value, as we shall show, stems from its emphasis on environment, economy and equity as three intertwined policy realms that must be addressed if sustainability is not to be depreciated or glossed. So often sustainability is diluted or devalued in programs and policies by leaving out one of the spheres or legs, most often the social justice one. This helpful picture of sustainability as three overlapping spheres or a three-legged stool or tripod is also vitiated by claims that the policies representing these three orbs may be pursued in their own right and then balanced. For example, the excesses of economic growth can be repaired or counterbalanced by environmental and social policies that remedy the excesses (or what economists call the negative externalities) of the growth imperative.

Another starting point or anchor for defining sustainability or sustainable development grows out of the most popular, and even hegemonic, of understandings of these core concepts. That is the Brundtland Commission's interpretation. This definition was originally fashioned by the World

Commission on Environment and Development, known as the Brundtland Commission after its chair, Gro Harlem Brundtland. In its 1987 report, the commission stated that sustainable development is "that development that meets the needs of the present without compromising the ability of future generations to meet their own needs" (UN, World Commission on Environment and Development, 1987: 43). While acknowledging that sustainability embraces elements of this rendering (e.g., the idea of generativity – or what the present generation owes to future generations), we believe its vagueness and generality has both contributed to its popularity and fomented great confusion about this core concept's components and requirements. (For greater elaboration, see Chapters 1 and 3.)

Our approach, by contrast, will challenge the essentially contested nature of sustainability by offering a series of definitions and interpretations of sustainability that will lead, we argue, to a strong and robust operational definition of sustainability working at the scale of the city. That is, we will opt for strong sustainability over weak sustainability as our standard for policy formulation and political struggle by politicians, legislators, grassroots, other nongovernmental organizations (NGOs) and the sustainable movement generally. Such a definition interprets sustainability as a *local, informed, participatory, balance-seeking process, operating within its Sustainable Area Budget, exporting no harmful imbalances beyond its territory or into the future, and in so doing opens spaces of opportunity and possibility* (Dumreicher, Levine, Yanarella and Radmard, 2000: 290). We call the preceding an operational definition because it affords researchers, policymakers and other change agents the means for understanding sustainability in a manner that provides a sustainability yardstick for measuring and assessing sustainability that goes beyond conventional metrics, like sustainability indicators pursued, green jobs or businesses created or even ecological or carbon footprints.

This alternative strategy or theory of sustainable cities requires an appreciation of unconventional concepts like strong urban sustainability, Sustainable Area Budget, the Sustainable City Game and rural partnerland, which are the four operating principles of sustainability that ultimately contribute to and give substantive meaning to this operational definition of sustainability. The effort to understand fully what this new theory or strategy entails by grasping the import of these nonstandard categories and ideas comprising our operational definition will reward readers with many fruits. At the very least, it will invite sustainability advocates and critics alike to see the pursuit of sustainability as both more difficult and more comprehensible than current views and approaches informing much of sustainability policymaking these days. It will also demonstrate why "picking the low-hanging fruit" – by tackling the easiest and most inexpensive policies and programs to make the world less unsustainable – carries with it mounting costs and sources of opposition that can only be overcome by *not* picking the low-hanging fruit, but instead striving to make the world more sustainable by choosing an alternate process limned in the strategy of sustainable cities (or better city-regions) presented in this book.

The Argument Summarized: Strategy, Concepts, Cautions and Case Studies

So what does the overarching argument of this book look like? We will outline its structure, framework and general substance.

Part I of the book introduces a set of strategic and theoretical considerations influencing the sustainability debate and the questions of the proper scale for realizing the goal of global sustainability. Issues of city and countryside, technological innovation, the role and importance of local knowledge and incremental steps versus holistic design – these and other issues are openly raised and carefully addressed.

In Chapter 1, we begin at the global level, asking whether diverse strategies of sustainable development pursued at the global scale by international organizations like the World Bank and advocated by leading environmental institutes like Worldwatch lead to sustainability. The shortcomings and contradictions of the present global-scaled theory and practice of sustainable development as concept and strategy suggest that the very idea of sustainable development must be subjected to the most thoroughgoing reevaluation. This leads to a trenchant critique of these efforts and issues in a negative answer to this query. Working off the adage popularized by Hazel Henderson, "Think Globally, Act Locally" (1988: 354), this chapter endeavors to show that thinking globally and acting locally provide an alternative approach to sustainability that can best be understood by recognizing the importance of relevant lessons from the sustainability and natural ecosystems (also called "climax systems") for achieving social sustainability. It also draws out the implications of "going local" for grassroots sustainability groups while simultaneously fashioning nonhierarchical networks across the globe that are dedicated to publicizing successful projects and promoting sustainability education, environmental improvement and worldwide stewardship of scarce, depletable resources by local nodal points. In rejecting simplistic versions of sustainable development approaches, this chapter offers the strategy of sustainable cities both as an alternative strategy and as a catalyst to long-term global sustainability. In so doing, it does not seek to replace all components of the strategy of sustainable development, but instead to place them within a restructured strategic framework with the design and building of sustainable cities at its center.

Many avant-garde groups and movements within art, architecture and politics populating Western society have found it necessary to present a plan, a manifesto, of their own new and iconoclastic way of representing, building and acting upon the world. Chapter 2 offers a sustainable cities manifesto that performs a similar service to the sustainability movement. It is divided into three sections – pre-text, text and post-text – in order to outline the historical and theoretical origins of the future sustainable city; the core ideas animating the project of sustainable cities; the future unfolding of this project, which is built upon new ideas, novel techniques and technologies; and new relationships between the built environment and the Earth's ecosystem. We begin by tracing the origins of the twenty-first century sustainable city to the Greek polis and the medieval commune (specifically the Italian hilltown); then we seek to recover convergent ideas and aspirations in modern efforts embodied in Howard's garden city movement, Geddes' Outlook Tower and organic town planning approach and Mumford's regional planning association program in order to bring urban planning into scale with the requirements of the values of true urbanity and in line with the imperatives of the ecosystem.

This chapter then identifies the core principles of a strategy of sustainable cities (actually sustainable city-regions) opened up by these historic examples and their urban heritage, and emerging from the global ecological crisis that compel the contemporary research university, eco-scientific praxis, and democratic policymaking to respond to the radical implications of this crisis by shifting cities and city-regions onto truly sustainable foundations.

The burden of Chapter 3 is to introduce a wide audience to the state of the sustainability debate today – a dispute that has attracted some of the cleverest minds and most successful sustainability advocates in marketing ecological practices and technologies. We argue at the outset of this chapter that despite the widespread involvement of noted ecologists, eco-scientists, policy analysts, entrepreneurs and even politicos, sustainability remains an "essentially contested concept" and sustainability discourse a continuing cacophony. The latest expression of this Babel of conflicting voices is evidenced in the emergence of "green-talk" and growing substitution of varieties of "greenness" for sustainability and sustainable development in everyday and media parlance. In this introduction, we seek, first of all, to compare and draw out differences between the rising tide of green products and program promotion and the active pursuit of genuine sustainability. If joining what Tom Friedman has called a "green party" (2008: 205) allows largely piecemeal, incremental environmental improvements to be undertaken anywhere, we seek, secondly, to show why the massive shift to the city from the countryside across the globe calls for a more careful cataloguing and accounting of the many strains of so-called sustainability strategies. To do this, we offer a hierarchical sustainability rubric that allows us meaningfully to offer gradations on the sustainability continuum that suggest the city as the proper scale for instituting strong sustainability on *terra firma*. Along the way, this chapter illuminates the stakes involved in this conceptual debate and provides greater clarity about what these putative variations on sustainability imply for both local-global theory and practice.

As the ecological movement shifts from conceptualization to implementation of programs for sustainability, the matter of appropriate tactics and strategy comes into play with increasing frequency. This strategic and tactical review becomes all the more crucial in the twenty-first century context of global warming and the peaking of world oil reserves. In Chapter 4 we argue that the currently fashionable and appealing tactic of "picking the low-hanging fruit" in order to achieve quick paybacks and build coalitions of support for sustainable policies is fundamentally flawed and, on its own logic, counterproductive. Grounded in the imagery of a pathway to sustainability, this policy nostrum eventually leads to a series of barriers that make the goal of that journey unrealizable. In advancing the alternative strategy of sustainable city-regions, this chapter lays out the case for adopting the commitments, tactics and stratagems flowing from the idea that sustainability is a balance-seeking process requiring the establishment of the minimum level of activity that would make each succeeding step easier, not more difficult. It also assesses the continuing arguments supporting the low-hanging fruit dictum and responds to the political realities that often make this simplistic approach seem compelling.

Chapter 5 begins with a catalogue of diverse sustainable city programs that have recently been initiated in the United States, Canada, Europe, China and the Middle East; we also note that many of these projects have largely remained contained within the boundaries of the city or metropolitan area, giving scant attention to the wider region and its relationship to the city's ambitions to achieve sustainability. In moving from the city to the city-region as the smallest unit within which the many imbalances constituting the twenty-first century built environment can be rebalanced, this chapter highlights the crucial importance of defining a yardstick of sustainability to measure the city-region's rightful share of the Earth's resources without extending its ecological footprint beyond the bounds constituting what Wackernagel and Rees (originators of ecological footprint analysis) call the city's "fair Earthshare" (1998: 53) – and we call its Sustainable Area Budget (SAB). After offering a review of the evolution of the SAB, we present a clear and detailed method of calculating the SAB, acknowledging the existence of continuing questions and complexities in generating an accurate calculus of it. The chapter also underlines the compelling rationale for the SAB and contrasts its virtues with the shortcomings of the more popular Sustainability Indicators approach. This chapter closes with an introduction to the idea of a partnerland as a necessary complement to the city. The partnerland concept represents a small, but essential, step in the reconstruction of the global economy on a sustainable basis. The basic operating goal of any sustainability regime is to assure that the movements of materials, energy, goods and services are activated by the larger, more comprehensive forces of supply and demand that include ecology and equity as balancing influences on traditional economic forces. The partnerland concept is a novel way of implementing a first operational step at the scale of the city-region. We close this chapter with an overview of the architecture of the Sustainability Engine, showing its central importance to the advance of an urban sustainability agenda and highlighting its precursors.

Next, Chapter 6 introduces the powerful metaphor of a game into the theory and practice of sustainable cities. Once the SAB has been formulated and fully operationalized as a means of defining the sustainability midpoint of a city-region, a subtle and robust research program presents itself and a kind of "Sustainable City Game" can come into play. The Sustainable City Game is conceived as both a pedagogical exercise and a policy tool that could be played for educating policymakers and training citizens in the practice of generating what we call sustainable town or city implantations in an existing city-region. As such, it lays the basis for integrating a policymaking process in an existing city-region. With the assistance of an array of new software tools on the shelf or on the horizon, the local-regional policymakers and other stakeholders of a given city-region could negotiate amongst themselves how they would afford to live within the limits of their land-based budget through their own creativity and ingenuity. In this new governance process, which involves an expanded democratic process, participants would consider their own traditions, land, climate, history, hydrology, agricultural traditions, land resources, culture, patterns of conviviality, urban and rural patterns, recreation and resources, etc. in the process of projecting alternative scenarios for their collective future. In this context, the role of the Sustainability Engine in democratic policymaking becomes obvious. That is, in the Sustainable City Game, various parties would be free to propose any policy alternative they choose, however self-interested or self-serving. In principle, everything could be negotiated, except sustainability itself. The engine would become the feedback tool for evaluating the state of ecological balance of these various scenarios. It would also serve as the resource for assisting the player stakeholders in balancing city scenarios emerging from the different proposals put on the table. As the game is played, researchers responsible for developing the engine would assist in modeling the new scenarios and in developing new modules to support these expanding experiments. The advantages over traditional policymaking and the benefits flowing from overcoming narrow self-interests (that originally anchor playing the game) are carefully and, we hope, compellingly enumerated.

Having illuminated the conceptual, theoretical and strategic foundations of our approach, Part II gathers a number of case studies drawn from prominent initiatives in real world sustainable city projects and investigations, as well as projects undertaken by the coauthors and their partners.

Chapter 7 explores the strategic implications of linguistic clarification of key terms, i.e., the "local" and the "global," in the sustainability debate triggered by key participants in that policy contest (Wendell Berry, Alan AtKisson and Wolfgang Sachs). Then, we offer a critique of the deployment of those terms in the strategy of "green globalism," which is a program identified with the policies of many nations of the north and of those international organizations dominated by industrial nations. This critique morphs in the second section into a critical examination of the perils of environmental moralism expressed in the proliferating number of international sustainability declarations enunciated at international conferences and multilateral meetings. In particular, we contrast the favored strategy of sustainable development growing out of the local implementation of the imperatives of Chapter 28 of Agenda 21 and its call for local authorities to put into practice at the city or municipal level the principles underpinning this document, which emanated from the Rio Earth Summit with the Aalborg Charter (a document that the coauthors directly contributed to writing and that serves as the basis for several hundred programs among European towns and cities today). In the third section, we critically examine four representative sustainable community programs in various parts of the world to determine the extent to which the tacit model of sustainable development championed by the Local Agenda 21 process has borne fruit. This chapter concludes by speculating on the daunting task ahead for advocates of the city as the fulcrum of global sustainability as they seek to overcome today's unequal power relations that heavily shape the global political economy and influence the character and direction of north–south relations.

Chapter 8 uses a case study of Chattanooga, Tennessee to highlight a typically American approach to sustainable cities. In 1969, Chattanooga had the dubious honor of being dubbed the most polluted city in the United States. Prominent political leaders, the local chamber of commerce, other key community stakeholders and a major regional foundation then set in motion a Sustainable Chattanooga program dedicated to overcoming this blight and rebuilding this southern city along sustainability lines. The result, as we show, was an example of a corporate-led program that constructed a form of Showcase Sustainability, which revitalized the downtown and significantly reduced local pollution but did not take into consideration significant social equity issues that would comprise the third leg of the so-called sustainability tripod (economic health, environmental protection and social equity). Using the strong sustainability framework outlined in Part I of the book, this chapter delves into the underlying reasons for the failure of this success story to institutionalize authentic sustainability at the city-region scale. It also points to why subsequent groups dedicated to sustainability in the Chattanooga area have had to coalesce after Sustainable Chattanooga supposedly completed its work in order to press anew with the local-regional effort to realize the sustainability dream.

Although Canada and the United States have long been regarded as friendly neighbors in an uneven relationship where Canada has been seen as the weaker and more docile partner, the case study of the town sustainability program in Okotoks, Alberta detailed in Chapter 9 discloses how Canadian efforts in environmental and sustainability policymaking have actually vaulted ahead of the United States. This chapter sketches out the way that municipal policymaking and governance in a small Canadian town took advantage of distinctively Canadian cultural and political values and processes to develop a model program that works from a sustainability yardstick – in this case, the upper population limits of the Sheep River's carrying capacity – to organize a town sustainability program. Combining top-down commitment with bottom-up embrace of strong sustainability, the Okotoks case, as we demonstrate, reminds us that sustainability city making is an ineluctably political enterprise. The chapter closes with some lessons to be drawn from this case study and offers a few caveats about this Canadian municipal program's oversights and weaknesses, as well as mounting threats from Calgary, the metropolis to the north.

In many respects, European cities are more fortunate than American ones in having a number of key cultural and other assets, e.g., denser and more contained urban forms, distinct surrounding agricultural regions and greater appreciation of urbanity's values, which allow them to generate significant and robust sustainable city-region projects. Chapter 10 identifies one that took advantage of those native benefits and showed great promise. The focus of this chapter is on Vienna's Westbahnhof project, an ambitious undertaking that completed three phases of its conceptualization and design at the coauthor's Center for Sustainable Cities at the University of Kentucky, in Lexington, Kentucky, in association with Oikodrom, the Vienna Institute for Urban Sustainability in Vienna, Austria. Because the underlying framework of this project departed in significant respects from conventional approaches to sustainability, the chapter begins with an extended outline of its theoretical underpinnings and design elements, as well as the historical model and precursor to its "city-as-a-hill" design – the Italian medieval hilltown.

Then, through text and illustrations, we explore the fundamental effort to integrate urban architectural design with strong sustainability principles into a program for implementing a sustainable city implantation (SCI) upon and over the current Westbahnhof railway terminal site in Vienna. The purpose is to demonstrate not only that sustainability in a place is imperative, but that it is feasible. Like so many innovative and unconventional efforts, the Westbahnhof project remains unbuilt. Had it been completed, this sustainable urban implantation would have become the first project in Europe to fully implement the principles of the Aalborg Charter and, in so doing, become Europe's first modern sustainable city-region.

In Chapter 11, we shift attention to the massive continent of Asia, where China, the sleeping giant, slumbers no more. Even as its city-building programs moved with relentless speed to construct a projected 300 new cities between 2005 and 2010, China has looked inwards and outwards for experts with ideas for fostering economic and cultural development within its rural countryside. One such program was Oikodrom's Sustainable User Concepts for China Engaging Scientific Scenarios (SUCCESS) project, a recently concluded five-year interdisciplinary study supported by a three million dollar research grant from the European Union. Working from the dual concerns of what to maintain and what to change in a top-down and bottom-up framework, several dozen natural scientists and social scientists investigated the proto-sustainable possibilities of seven Chinese villages in six different provinces. With the many local-national and global policies and pressures threatening the continuing existence of China's hundreds of thousands of rural agricultural villages in anything resembling their current form, we show in this chapter how the SUCCESS project was aimed at exploring a variety of future options for these villages, with an emphasis on those scenarios that promised to evolve into sustainable future settlements. This chapter clarifies how historically and culturally grounded local knowledge of villagers uncovered by this project could be integrated with Western expert knowledge in a democratic process that yielded important findings, while generating intercultural goodwill. Although China continues along the path of urbanization through massive city-building projects, the lessons of this project point to the continuing viability and possible evolution and transition of China's village into larger and more viable sustainable town-country settlements (or better sustainable city-regions) with greater diversity and life choices than those currently available in small rural farming villages.

Amidst the growing popularity of the village scale and specifically the eco-village as the model for rural development in China, Chapter 12 offers a case study of a showcase eco-village plan quite dissimilar in design and planning process from those featured in Chapter 11. It focuses upon Huangbaiyu, a small village in the eastern Liaoning province of China, and presents a critical case study of this village's aspirations as an object lesson of how a particular understanding of sustainability developed by William McDonough and Michael Braungart and promoted as a collaborative partnership among its foreign designers, the Chinese national government, an NGO, a local developer, a major university and a local planning group led to its precipitate and abysmal failure when implemented. Our major concerns are to answer three questions: How successful was the McDonough and Braungart model of "cradle-to-cradle" (C2C) approach in Huangbaiyu? What were the sources of its untimely and abject failure? What does Huangbaiyu's fate say about the state of sustainability planning in China? Using the critical strategic framework of strong urban sustainability outlined in Part I, this chapter locates a key source of the failure of this recent Western-directed project in the absence of any serious effort to institute a strong democratic consultation and participation process fully involving the local villagers in what was billed as a model eco-village initiative intended to be replicated throughout China.

In the Conclusion, we close with a summary of the road most traveled to date in projects and programs designed to build sustainable cities and city-regions. We argue that only the road less traveled – i.e., the strong sustainability strategy outlined in this book and used as a critical evaluative framework for analyzing representative cases around the world – holds real promise of creating sustainable city-regions guided by a toolkit of ecological and technological conviviality. Only this approach offers hope of closing the gap and overcoming the frictions between the nations of the north and those of the south by greening the north and providing exemplary strategies, tools and programs for the developing nations of the south to pursue an alternate and ecologically resilient path to true sustainable development. Only then will incrementalist, pie-in-the-sky and mythical notions and strategies, which are being advanced in the sustainability debate, be surpassed by the achievement of true sustainability locally-regionally-nationally and globally. Only then can many of the deeper causes of international war, conflict, and unrest be mitigated or overcome.

Part I

STRATEGIC CONSIDERATIONS

Chapter 1

DOES SUSTAINABLE DEVELOPMENT LEAD TO SUSTAINABILITY?

Despite the growing competitiveness in the international political economy for new and more tightly integrated global markets and the rise of new and ascendant economic powers in the Orient, the need for sustainable development on a worldwide scale remains a pressing one on the international agenda. In thinking schematically about the trends and tendencies influencing the institutionalization of sustainable development, an image of where we are and where we want to go may be represented by a series of parallel forces moving us toward imbalance or collapse intersected by a series of parallel forces moving in the opposite direction toward balance or sustainability. One model of desired change suggests that if we can only build up the arguments, tools, and resources for the sustainable direction, then the balance will shift at some point and we will then be effectively on a sustainable path. Another model of desired change, the sustainable city alternative, has historically been the road less traveled.

These sustainable tendencies, we argue, can at best provide certain necessary conditions for generating long-term sustainability – strategically, they cannot constitute the path itself. *Sustainable development alone does not lead to sustainability*. Indeed, it may in fact support the longevity of the unsustainable path. These many tendencies require synthesis and transformation into a new system of production and reproduction that can best come from an anchor and a catalyst – i.e., the design and institution of sustainable cities. In our view, sustainable cities act as necessary nodal points for the transformation of quantitative change into qualitative change.

To support this counterhypothesis, we undertake an immanent (i.e., internal) critique of the underlying assumptions grounding the strategies of sustainable development advanced by some of its most prominent and articulate advocates. In critical social inquiry, the idea of immanent critique was most forcefully and effectively applied by Karl Marx in his analysis of capital. This approach to critical analysis seeks to demonstrate how within the framework of assumptions, beliefs and arguments grounding a particular theory or position certain contradictions arise to compromise the integrity of the stance, thereby opening up analysis to amendment and perhaps wholesale overturning. In challenging the internal logic of dominant strategies of sustainable development, it is not our aim to unseat the goal of sustainable development; rather, we intend to show why some of its guiding assumptions and key tenets are inconsistent with each other or with its deeper values. We also intend to demonstrate how an alternative political grounding or approach – in some cases, faintly limned within the dominant strategy – might overcome these deficiencies and contradictions.

In what follows, we review and criticize the global path or strategy of sustainable development that was the dominant intellectual and policy trend and orthodoxy in policymaking communities and world organizations and forums in the decade of the nineties.

The Meaning of Sustainable Development

Since the mid-1960s, the literature on sustainable development has grown and proliferated. This growth has been inspired in large part by a multitude of individual and group initiatives in sustainable living and design from John Todd's New Alchemy Institute to Paolo Soleri's Arcosanti. It has been no less stimulated by the broad spreading out of ecological consciousness and of the globalist framework informing it. Futurists, like Hazel Henderson and William Irwin Thompson; scientists, like James Lovelock and Lynn Margulis; philosophers, like the late E. F. Schumacher and Arne Naess; social ecologists, like Murray Bookchin and Ynestra King; and solar energy/conservation advocates, like Amory Lovins and Paul Hawken along with other renowned figures have comprised the key figures of the broad movement weaving together alternative technology, soft energy paths, economic decentralism, local democracy and an ecological vision. All have contributed greatly to the outlines of a vision of global sustainable development and a generally convergent set of strategies and tactics for reaching that future condition.

Dialectical fealty to the lofty goal of global sustainable development requires us, however, to raise the question whether this unfolding vision and strategy may not have congealed into a kind of orthodoxy that inhibits as much as fosters the transition to a condition of ecological and social sustainability. The first section of this chapter grapples with the meaning of sustainable development as it appears in recent theoretical and programmatic statements by some of its chief expositors. In the course of critical review, our own approach to sustainability is presented. This exercise is intended both to clarify the substance of this key term and to position our own rendering within the range of meanings found within the literature.

For much of the late nineties until the turn toward sustainable cities and communities, the objective of sustainability had been appropriated and thematized primarily in the international development literature. We should therefore expect that term and its partner, development, to receive careful discussion and precise elaboration. Yet, for a concept so central to the abovementioned literature, the meaning of sustainable development remains quite elusive. Despite the myriad writings celebrating this idea, precious little intellectual work has been expended in defining its substantive core or formal boundaries. In a survey of many of the best works in the field during that time, its meaning is either taken for granted and therefore unexplicated or is defined very loosely, and then taken as a point of departure for outlining the shape of a future sustainable society. Yet, within these limits and with these shortcomings, it is possible to draw out its broad parameters and some of its shared elements, as well as several issues and dimensions which remain contested terrain even among its chief promoters.

Perhaps the most comprehensive and detailed portrait of the contours of sustainable development and its goal of global sustainability has been offered by Lester Brown (1981). Setting his conception of sustainability against the backdrop of those destructive and burdensome trends emanating from industrialization and economic modernization, Brown defines the sustainable society in terms of population stabilization, the preservation of the Earth's renewable natural resource base, the prudent use of land (urban, farm, and wilderness), the protection of biological systems (including fisheries, grasslands, and forests), the reforestation of the Earth, the maintenance of the web of life from microorganisms to the gene pool to extant species and flora and fauna, and the conservation of energy resources through the shift to renewable sources of energy.

What is surprising about this delineation of sustainability is that this treatment nowhere specifically defines its core meaning. Instead, Brown offers at best a wide-ranging, but not exhaustive, enumeration of the constituents of societal or global sustainability. Moreover, as Paehlke (1989: 66) notes, "given Brown's middle-of-the-road position on environmentalism, Brown's vision of a future sustainable world or society is underpinned by a commitment to slowed, but continuing, economic growth." In sharing this vision with Brown, two other sustainable future advocates, Robert Hamrin (1983) and Dennis Pirages (1977), make clear that this position amounts to a commitment to sustainable *growth* rather than to sustainable *development.* For example, Hamrin makes this point explicit when he notes: "The concept of sustainable growth encompasses the necessity of continued economic growth to meet basic human needs around the globe while also underscoring that the growth must be of such a nature that it can be sustained indefinitely by respecting nature's boundary conditions" (Hamrin, 1983: 7). Likewise, in calling for the social design of sustainable futures on a societal and worldwide scale, Pirages lays out an agenda for sustainable growth defined in terms of "economic growth that can be supported by physical and social environments for the foreseeable future" (Pirages, 1977: 10).

As Samir Amin (1976) – one of the most renowned dependency theorists of the 1960s – observed some years ago, the quarrel over the difference between growth and development is no mere semantic disagreement. For Amin and other dependency theorists, growth is not development and this distinction becomes theoretically and strategically important in peripheral countries where economic growth can mask the absence of true economic development, which has been blocked by relations of dependency with core or metropolitan nations of the northern hemisphere. More recently, Daly and Cobb have recontextualized this point within the sustainability debate by defining "growth" as "quantitative expansion in the scale of the physical dimensions of the economic system," while defining "development" as "qualitative change of a physically nongrowing economic system in dynamic equilibrium with the environment" (Daly and Cobb, 1989: 71). As they argue, "by this definition, the Earth is not growing but developing. Any physical subsystem of a finite and nongrowing Earth must itself also eventually become nongrowing. Therefore, growth will become unsustainable eventually and the term 'sustainable growth' would then be self-contradictory. But sustainable development does not become self-contradictory" (71–2).

Part of their argument against versions of sustainable growth is grounded in their effort to find an alternative to gross national product (GNP) and net national product (NNP) as a measure of sustainable development and in their defense of "strong sustainability" over "weak sustainability" as the core element of sustainable development. Although the elaboration of the former gets into some complicated and subtle economic argumentation, which need not detain us here, the latter distinction carries our analysis of the key concept of sustainability another step further. Against the assumption of the "human exceptionalist" school headed by Julian Simon (1981) and the late Herman Kahn (1976) that postulates the infinite replaceability of finite natural resources by humanly created substitutes, Daly and Cobb press their defense of strong sustainability by arguing that the latter requires "maintaining both humanly created and natural (resource) capital intact separately, on the assumption that they are complements rather than substitutes in most production functions" (Daly and Cobb, 1989: 72). The implications of this argument have been concretized by Amory Lovins (1977) in his advocacy of soft energy paths that draw on renewable energy income instead of hard technology paths that exact a heavy price in terms of scarce human capital.

Within the international development literature, clarification of the sustainable development concept has been advanced in the work of Michael Redclift (1987). Working from the proposition that sustainable development socially can be discerned in part by working from the dynamics of ecosystems, Redclift shows how complex ecosystems, like the tropical rainforests, achieve eco-sustainability or homeostatic balance, or what natural ecologists call "climax systems" of high diversity, large biomass and high stability through protection from rapid change and "through shifts of energy flows away from production and towards the maintenance of the systems itself" (1987: 18). By contrast, he notes, human settlements typically seek to stall such ecosystems in early stages of ecological succession, where the yield of products is high, but where the stabilizing elements of organic matter and biomass fail to accumulate. High production within these ecosystems, then, comes "at the cost of confounding nature's strategy of maximum protection or adaptation" (18).

This image of *ecological sustainability*, we argue, is the model on which sustainable development should be patterned. Of course, natural ecosystems are not equivalent to social systems. Nor are the two ultimately separable. Because the relationship between human settlements and nature is one of identity and difference, human beings in communities live *in* nature, not with or against nature. Simultaneously, human beings in their nonidentity with nature have the capacity of reason to shape their environment and to develop tools to alter or to adapt to their changing environment. Moreover, larger social relations and human groupings (organizations, cities, governments, economies) in their social and historical configurations can strongly influence the compatibility and convergence between natural and social systems.

These philosophical points can be made more concrete by considering other arguments put forth by Redclift and elaborated by others. A concept of sustainable development grounded on the sustainability of natural ecosystems, as Redclift suggests, implies a subtle appreciation of the relationship between the carrying capacity of the ecosystem and human population growth. "The concept of 'sustainability' makes little sense," he says, "unless we also consider the impact of rapid population growth on the physical resource base" (1987: 29). Carrying capacity is a term engendering much controversy. Daly and Cobb (1989) note that some question its appropriateness to human beings. As they acknowledge, the carrying capacity of the environment for animals can be calculated basically in terms of population, since "per capita resource consumption for animals is both constant over time (animals do not experience economic development), and constant across individual members of the species (animals do not have rich and poor social classes)" (136). Moreover, technology for other animal species "is a genetic constant, while for humans it is a cultural variable" (136). Yet, sustainable development theorists continue to

Table 1.1. Competing paradigms

	Industrial-technological paradigm	Environmentalist paradigm	Ecological paradigm
Nature	Natural resource – object of exploitation for human gratification	Natural resource – object of domination moderated by technical constraints and environmental limits; Earth as a spaceship; resource efficiency and conservation	Humankind's organic body; Earth as Gaia
Environment	Air, water, soil, space – a "free good" or dumping ground for "negative externalities" (pollution, waste) of the industrial production process	Primeval nature; wilderness untouched by human industry	Biosphere/ecosystem – the life sustaining natural processes of which human life is a part
Science	An organized system of knowing intended to rationalize and control nature and an instrument for advancing R&D to achieve long-term corporate goals of market control and profit	A means of understanding the negative consequences of human intervention into nature and a political resource for defending the environment against social-industrial assaults	An ecologically grounded science purged of scientific instrumentalism and technological domination
Technology	Capital-intensive, complex technological forms undergoing constant innovation	Attachment to advancement of high technology and to use of antipollution devices as technological fixes	Appropriate technology; energy nonintensive, labour intensive, forms of technology fostering political and economic decentralization
Politics	An artificial system for mediating competing interests and pressures where corporate power is applied to access points to exert political influence and to shape the public agenda	Interest-group politics and legalism	A new politics beyond mere reformism and interest-group accommodation guided by a postmodern cultural consciousness and restructured social or institutional vision

State	Class-based statement for socializing risks and privatizing profit	An agency for promulgating and enforcing environmental laws and policies	A medium for redesigning policy toward biospherically sustainable production, for redirecting public capital toward low-energy, labour-intensive, appropriate technology, and for redistributing the social surplus toward social welfare qua inclusive social services.
Economics	A techno-corporate political economy strapped to the treadmill of production with no limits to growth	A postindustrial political economy balancing economic production and environmental safeguards guided by the strategy of internalizing the negative externalities of the production process	A post-Keynesian, steady-state economy oriented toward simpler human needs and maximizing risk aversion.
Human Nature	*Homo economicus*/possessive individual/consumer self	Liberal self constrained by limited social enlightenment and environmental awareness	A posthumanist, decentered conception of the social self entwined in a more encompassing framework of natural processes
Good society	An economically concentrated, class-stratified society geared to infinitely expanding human wants committed to market expansion and planned obsolescence	A New Age society combining high technology and environmental protection, biased toward middle- and upper middle-class, professional values	A democratically planned, decentralised society promoting political participation, ecological sustainability, meaningful work, and elegant frugality

Source: Brunn and Yanarella, 1987

argue with much persuasiveness that, despite these differences, carrying capacity remains relevant to the human population.

For one thing, because we live in nature, the overburdening of the limits of other ecosystems (water, forest, grassland, etc.) by human demands and ecological damage endangers long-term human survival (Odum, 1989). Second, the claim by proponents of the "human exceptionalist" school that natural resource limits can be infinitely replaced by technological substitutes remains simply an article of faith badly shaken by closely argued critiques of that position (Ophuls, 1977). Finally, the alternative assumption of latter-day ecologists of every ideological conviction – the presumption of ecological scarcity – suggests that "instead of simple Malthusian overpopulation and famine, we must now also worry about shortages of the vast array of energy and mineral resources necessary to keep the engines of industrial production running about pollution and other limits of tolerance in natural systems, about such physical constraints as the laws of thermodynamics, about complex problems of planning and administration, and about a host of other factors Malthus never dreamed of" (Ophuls, 1977: 9). "If ecological scarcity is conceived as this ensemble of separate but interacting limits and constraints on human action" (9), then its essential meaning is that humankind's political, economic, and social life "must once again become thoroughly rooted in the physical realities of the biosphere" (133).

Still, even if ecological scarcity overwhelms the ability of human ingenuity and industry to expand the capacity of the ecosystem to carry an ever larger human population at a decent quality of life, the problem of overpopulation for the world today in terms of sustainability, as Redclift argues, "is not so much the net increase in population at the global level, but the rate of change in population in the most critical regions" (Redclift, 1987: 30). Context-specific analysis, not global doomsaying *a là* the Ehrlichs, is what is required. Clearly, crucial to ecological sustainability in those areas of the world experiencing the highest rates of population growth are policies fostering "environmental security" – i.e., practices that preserve indigenous natural resources and prevent environmental degradation (Redclift, 1987: 32).

But, as Redclift's book makes evident, the difficulties of those countries of the south in achieving environmental security and promoting sustainable development programs originate for the most part in models of industrial and agricultural modernization and programs like the "Green Revolution" (Henderson, 1988) imported from the north. These programs are perpetuated by the finite nature of the energy subsidy provided by depletable oil and natural gas supplies for our wasteful and environmentally destructive forms of modern industrial and agricultural production. More generally, the continuing impact of relations of domination and subordination within an unfolding global political economy poses some of the most daunting problems for moving toward global sustainability.

Although many economists have not been persuaded that the environmental crisis presents an unsolvable anomaly to the paradigm of neoclassical economics, the alternative ecological paradigm informing sustainable development strategies presumes that "the lessons of ecology can, and should, be applied to economic processes" (Redclift, 1987: 32–3). This alternative paradigm is contrasted with two other competing paradigms evidenced in the local, regional, national and global development debates. This table serves to fill out philosophical bases for our conception of sustainability by comparing the images of grounding concepts of each paradigm and simultaneously posing the stakes in the intellectual and cultural controversies over economic development strategies and ultimately different global futures.

For us, then, sustainable development is patterned after the model of ecological sustainability guiding mature ecosystems at the stage of climax. Simultaneously, we recognize that the differences between human beings and other animal species and between natural and social systems pose different problems for human populations while permitting human rationality and socially constructed technology to offer limited means for extending the carrying capacity of the ecosystem of which humans are a part. Still, in embracing the emergent ecological paradigm outlined above, we are avowing our commitment to those widening visions of the future of the human species inspiring the cultural consciousness and sensibilities of diverse green movements throughout the globe.

An Immanent Critique of Sustainable Development Strategy

Granted that sustainable development is an essential tool for taking human development beyond the petroleum age and for restructuring relations between the north and the south on a more equitable basis, by what strategies and tactics might this objective be reached? Numerous transitional strategies have been offered and most of them have been grounded in a set of common or overlapping assumptions about the scale of the effort, the means to be employed, the relationship between global and local activities and the role of the sciences and social sciences.

The Issue of Scale: Societal or Global?

Among the key proponents of sustainable development, the overwhelming majority pitch their strategies for sustainable development at the scale of the nation or even the entire globe. Without question, the globalism of futurists like Lester Brown, Hazel Henderson and others is one of their greatest analytic strengths and among their key contributions to theorizing about sustainable development in its many dimensions. Thinking holistically about the care and maintenance of the ecosystem and the human species illuminates the complex but integrated relationship between the destruction of the Amazon rainforest and the problem of global warming and between the hole in the ozone layer and the industrial and household use of certain chemical aerosols. On the other hand, the very globalizing vision of ecological thinking when transferred to strategizing can militate against the tacking down and institutionalization of self-sustaining processes at lower scales where sustainable development may be more realizable and the results more palpable.

Targeting sustainable development strategically at the global scale makes meaningful and concerted political action all but impossible. However interwoven the "community" of nations with the larger ecosystem and with the international political economy, the mix of sustainable and unsustainable tendencies across nations is often quite different; the comparative political resources for furthering programs and policies supportive of the politics of sustainability are varied; and the relative capacities of subglobal political and governmental groups, agencies, organizations and movements to structure and/or influence international bodies to advance the agenda of sustainable development are extremely differential.

The Means to be Employed: Technological Instruments or Idealistic Panaceas?

It should occasion no surprise, then, that globalist strategies of sustainable development oscillate between technocratic nostrums and idealistic and moralistic responses. Because the problems of the ecosystem are often portrayed in apocalyptic terms and the response to such daunting problems inspires a great sense of urgency, well-meaning sustainable development strategists sometimes revert to the language of technocratic planning and administration. That is, they speak of the need for global ecological planners in international agencies who must work with national political elites and multinational corporate leaders to manage these environmental crises (Kraft, 1977: 191). The problem is that technocratic designs and strategies only duplicate the social and organizational forms and mechanisms that helped to produce the ecological condition confronting human beings in the first place. If poetic license may be allowed to make a complicated political and theoretical point, ecological thinking is best rendered by reminding us all that it is not that we don't have enough time but that we only have time to bring into being the changes necessary to unweave and reweave our practices and relations with nature and others into a durable and sustainable cloth.

To the extent that sustainable-development-inspired planners were to move from global crisis to global crisis, using technological fixes to patch up larger structural problems, they would in effect tend to strengthen the systematic relations supporting unsustainability – especially when such "Band-Aid" solutions would lead to situations where these deeper ecological problems would fall below the threshold of public attention and the political momentum for more fundamental change would be allowed to dissipate. (For discussion of eco-technocratic solutions in science fiction works, see Yanarella, 2001.)

The other pole of the alarmist doomsaying type of ecological futurism is the idealistic call for a paradigmatic shift in cultural values and policy priorities to bring about the political conditions essential to realize the global agenda of sustainable development. Unfortunately, this call for cultural self-renewal easily lapses into a characteristically Western Protestant form of moralism that acts as a substitute for genuine political action (Lukacs, 1968; Ranulf, 1964). Too often, the prescription of a paradigmatic shift bypasses the political realm and misinterprets cultural values and unifying political symbols as individual phenomena whose change or transformation occurs cumulatively through individual acts of conversion (Taylor, 1971: 3–51).

This strategic mode fails because it is not a genuinely counterhegemonic strategy. That is, a more critical theory of late capitalist and existing and former state socialist societies interprets the shared bases of these systems or resource profligacy and depletion as dynamic systems that produce and reproduce conditions supportive of continuing waste and eventual exhaustion. Such a hegemonic theory also recognizes the need for concerted political will and action to overturn the underlying ideological supports legitimating those systems. By contrast, the many orthodox versions of sustainable development presume that incremental change over time will eventually reach a threshold that institutes a new system, which then produces qualitative change. In a sense, then, such strategies of sustainable development act as if the sociopolitical and cultural transformations necessary to inaugurate sustainable development have already taken place so that in the transition to the end-goal sustainability is paradoxically everywhere and nowhere. What is needed is a determination of the minimum scale or context at which given acts can effectively become part of a balance-seeking process, rather than the cause of a larger or more intractable problem elsewhere.

Indeed, gradualism *qua* reformism and paradigmatic change *qua* individual conversion may actually work against long-term sustainability by producing situations where sustainable-development-induced changes achieved in some areas relieve some of the pressures, which then allow nonsustainable development conditions to increase and even grow more pronounced in other places nationally or globally. In any case, given the elitist biases of many formulae for sustainable development, it is clear that many global futurists operating on an apocalyptic schedule of impending ecological doom conceive the shift from the modern industrial-technological paradigm to the ecological paradigm largely in terms of a change in elite attitudes and perceptions (Kraft, 1977). What this view misses is the depth of impact required of any truly counterhegemonic strategy organized around sustainable development, a new ecological worldview and the role of political education at every level – including the mutual political education of "teacher" and "student" – in advancing such a strategy.

The Relationship between Global and Local Activities

The response to these criticisms has been offered, in part, in the emphasis on global planning and local self-reliance by some sustainable development globalists like Hazel Henderson (1981: 354–405). Their strategic program is encapsulated in the edict, "Think Globally, Act Locally" (Henderson, 1988: 354). This stratagem is manifested in the many inchoate planetary groups and coalitions that have emerged since the decade of the 1960s to promote human rights, oppose civilian nuclear power and nuclear proliferation, advance soft energy paths and protect hunted and endangered species, etc. For Henderson, such new and increasingly effective organizations like Amnesty International, Greenpeace and Oxfam incorporate a number of elements crucial to a counterhegemonic agenda:

(1) they spread global consciousness of promoting ecological values, social justice and international peace;
(2) they serve as nonhierarchical exemplars of "how planetary coordination of vital functions such as environmental protection and wise stewardship of oceans, land, and mineral resources can be achieved through non-bureaucratic, democratic means" (Henderson, 1981: 387);
(3) they coordinate local and regional programs of action founded on mutually accepted global principles and nonviolent, but sometimes civilly disobedient, tactics; and
(4) they work to keep "production, consumption, and institutional participation in economic and political life as close to the local level as possible" (Henderson, 1981: 381).

Henderson's articulation of a sustainable development strategy based on this well-known and pithy dictum marks a considerable advance over other strategic formulae that pose the problem of sustainable development simply at the global scale. Moreover, her conception of nonhierarchical planetary coalitions and networks for diffusing worldwide political consciousness of ecological values and eco-development removes the issue of paradigmatic change from the individual to the intersubjective realm of broad-ranging collective thought and action. These coalitions and networks serve as anticipatory institutions of a global web of monitoring and coordinating agencies for preserving and sustaining the ecosystem.

Without wishing to diminish this budding worldwide constellation dedicated to eco-sustainability, we question whether such an organizational configuration can truly realize global sustainability. In our view, it may serve as an early warning system for monitoring vulnerable points in the complexly interwoven fabric of the ecosystem. It may launch local experiments and innovations in discrete forms of agricultural or economic sustainability or in forms of alternative technology. It may even go far in disseminating political consciousness and sensitivity to those new conceptions of nature, science, human need, economics and so forth comprising the elements of an incipient ecological worldview or paradigm. What it is not designed to do, however, is to find a place where sustainable development can be institutionalized in a coherent and concrete form. As we argue below, this crucial role is best fulfilled by a sustainable-cities strategy.

Role of the Academic Disciplines: Multidisciplinary or Interdisciplinary?

The attainment of sustainable development is an eminently political and organizational task, but it is no less an intellectual one. This point raises the issue of the role of the disciplines in strategies of sustainable development. Too often, sustainable development strategists take as a given the disciplinary bounds set up by academic communities and the narrow range of roles permitted by certain historically and socially developed definitions of professionalism and expertise. International conferences on problems of ecological sustainability and sustainable development typically bring together noted experts from the natural and social sciences, who then offer specialized knowledge pertinent to these global issues. Unfortunately, such multidisciplinary forums do not begin to address issues of global and ecological sustainability in any genuine *interdisciplinary* framework. The specialization of knowledge into discrete disciplines establishes substantive boundaries and narrowly bounded discourses that are protective of each discipline's fragile hold on an aspect of natural or social reality. On the other hand, ecology is truly a subversive science, as Paul Shepard (1982) and others have argued, since sustainability in its multidimensionality overflows narrow disciplinary perspectives and parameters. This is what makes ecology transgressive and sustainability subversive to efforts at disciplinary closure and disciplinary ways of articulating sustainability.

Redclift (1987: 45–51) has cogently argued that both neoclassical and Marxian economics are ill-equipped to advance the agenda of global sustainability. For him, both of these antagonistic forms of academic inquiry into the dynamics of capitalist economy remain stuck in certain nineteenth-century assumptions about technology as the instrument for harnessing nature, the productivist ethic, the market as allocator, the vices or virtues of central planning and the environment as an object of use. These antiquated beliefs and assumptions, he argues, must be supplanted by postcapitalist and post-Marxist formulations that take account of the interconnectedness and embeddedness of human society in natural ecology.

Similarly, Daly and Cobb (1989: 25–137) have attacked the misplaced concreteness and "disciplinolatry" of contemporary economics, arguing that neoclassical economic theory even in its modern guises misrepresents the individual, rational behavior, the market, land and economic success by a variety of means of theoretical abstraction and conceptual reification. In questioning these renderings of fundamental concepts in this discipline, they seek to reformulate its conceptual bases in a manner that relocates *homo economicus* as a real person existing within a web of concrete, social, political and ecological relationships. In rendering the real sociohistorical person as the intersecting point of a wide variety of relationships, they then develop an alternative economic model that gives moral priority to human communities within wider communities existing over time and tying present generations to a view of the economy as *oikonomia* (i.e., household economy), to responsibilities to future generations, and to the continuing health of the biosphere.

Hazel Henderson has also made her contribution to demystifying the narrow technical expertise of the disciplinary specialist and reconceptualizing the professional expert as knower *and* moral agent in the process of promoting sustainable development. She has called for a new ethics of the solar age that confronts the "anomalies of jet-set academic elites in first-class hotels at meetings purporting to address hunger and poverty" (Henderson, 1981: 360). In formulating her ethical conception of the professional as conscientized expert, she seeks to base that new professional role on the democratic statement of global agreements, manifestoes and declarations of principles and the rights and duties of all people. Moving well beyond token and self-policing ethical codes of various professional associations, Henderson sees these statements and declarations as a means for repositioning academic experts – within the painful realities and excruciating dilemmas of the global problems they are addressing – by involving them in a set of concrete responsibilities and obligations to confront those realities and dilemmas in their professional lives, social interactions and personal lifestyles.

An Alternative Strategy of Sustainable Cities: Five Operating Principles

Orthodox strategies of sustainable development, we have argued, are hobbled by a number of shortcomings and contradictions that vitiate the enterprise of achieving global sustainability. Their theory and practice are pitched at a global scale that obscures analytic connections with other levels and scales and dissipates political energies at so high a level that the concrete consequences of these actions remain diffuse and evanescent. These strategies too frequently vacillate between the pole of technocratic-instrumental practices and the pole of individualistic-moralistic actions, leaving the medium of political struggle and political education in their more intersubjective and collective representations undeveloped or underdeveloped. Even where strategies of global sustainability coordinate planetary activities and local initiatives, the location of concrete forms of sustainable development is absent. Finally, sustainable development orthodoxy still pays too little attention to the need to transgress disciplinary boundaries and recontextualize ecological analysis within more holistic and interdisciplinary frameworks.

Our answer to the title question of this chapter – does sustainable development lead to sustainability? – must be a provisional no. In going beyond orthodoxy, sustainable development strategies need sustainable cities as nodal or tacking points and models of coherent and integrated sustainable development, thus making such sustainability possible by placing it in a context within which it may be validated as a process. A strategy of sustainable cities addresses all of the pitfalls and inconsistencies, undercutting even more sophisticated strategies of sustainable development. As we argue, a strategy of sustainable cities has the virtue of: (1) shifting the sustainability quotient of sustainable development to the positive side by locating ecological and social sustainability within a place – i.e., the city; (2) overcoming the historical divisions and philosophical dualisms installed in the modern epoch by alienating and fragmenting processes of Western capitalism and industrialization; and (3) revisioning global sustainability within humanly scaled terms. Such a strategy is grounded in five regulative ideas or operating principles.

Table 1.2. "Five Operating Principles of Sustainability"

First operating principle	Individual and discrete programs of sustainable development do not necessarily lead to ecological or social sustainability
Second operating principle (the principle of homeostatic balance	A desire, activity or process can take its place in a larger system only by finding its balances with that more encompassing system
Third operating principle (the principle of chaotic excess or extravagance)	No natural process or social system can avert the action or intervention of those forces of chaos or disorder in the cosmos, because they are always already there
Fourth operating principle	All systems shall first be designed to seek their ecological balances, at the smallest scale, and in any event at no scale larger than the city/region itself
Fifth operating principle	All lower imbalances are to be negotiated upward (outward)

Source: Yanarella and Levine, 1992a.

The first such operating principle is that *individual and discrete programs of sustainable development do not necessarily lead to ecological or social sustainability.* The commonly held assumption that this principle calls into question underpins orthodox sustainable development strategies, as we have shown above, as well as the views of many activists within the environmental movement. Applying our immanent critique of sustainable development theories to mainstream environmentalist approaches, we argue that isolated, discrete actions promoting environmental repair have as great a probability of lowering the sustainability quotient (i.e., the overall or net impact on the dynamic interaction between unsustainable and sustainable practices, programs and processes) as they have of raising it. As such, programs like maximizing recycling or minimizing pollution, although they may contribute in a quantitative way to environmental improvement, may fail in a qualitative way to stimulate ecological sustainability.

Indeed, such efforts alone land outside a more encompassing synthetic framework and integrative-policy approach and tend to perpetuate the mindset which has caused these imbalances in the first place; they may even strengthen the tendencies toward unsustainability by promoting meliorist policy measures that simply postpone the ultimate day of reckoning with ecological limits. The political implication of this principle is not that environmental activists should withdraw from policy reforms. Rather, it is that they should seek tactics that promote "nonreformist reforms"

(Gorz, 1972: 7) and that locate these tactics and strategies within a more encompassing vision of ecological sustainability.

The second operational principle animating this alternative strategy is the *principle of homeostatic balance*. This suggests that activities or processes are neither inherently good nor bad when taken by themselves. Instead, a desired activity or process can take its place in a larger system only by finding its balances within that more encompassing system. In order to seek such a balance, the process must have a context or system within which the balance may occur. It is thus a question of relationships. For even when a new component introduced into an existing system of relationships upsets the balance of a larger system, a countertendency may be set in motion whereby the larger system responds to the change by striving for a new state of equilibrium. James Lovelock (1979, 2007) has made a strong case for conceiving Earth as a self-balancing organic system striving toward a steady state favorable to the maintenance of life and capable of adjusting within certain bounds to large-scale human interventions and industrial assaults. At the societal level, we argue, this larger balance-seeking system is the sustainable city.

Simultaneously, this balance-seeking principle must be supplemented by recognition of inevitable tendencies in the world and nature toward disorder and chaos.

This third principle – what deconstructionists might call the *principle of chaotic excess or extravagance – implies that no natural process or social system can avert the action or intervention of those forces of chaos and disorder in the cosmos because they are always already there.* Nature is indeed a plenum that overflows the homeostatic propensity of self-organizing systems in nature or society to restore order to living processes of growth and equilibrium. Likewise, increasingly complex social systems reside within a wider setting of contradictory natural forces and detotalizing social processes that exceed the capacity of human agency at whatever level to dominate and control these systems or processes without remainder. Yet, ironically, because this remainder, this excess, this supplement, is always already there and injects a salutary tension into any process striving toward balance and stasis, such chaos may be seen, as Prigogine and Stengers (1984) have argued, as the great creative force in the natural and social universe. Order-out-of-chaos is only achieved through the careful and continual negotiation of dynamic balance, through the lateral coupling of order and chaos and not through fruitless authoritarian-technocratic efforts to suppress or banish chaos from the world. In this respect, the highpoint of urbanity – the early mediaeval hilltown – represents the paradigmatic case of the achievement of dynamic social and political equilibrium without a dominating center (Mumford, 1961: 316).

The alternative strategy for sustainable cities also embraces a fourth regulative idea – namely, that *all systems shall first be designed to seek their ecological balances at the smallest appropriate scale, but in any event at no scale larger than the city itself.* That is, an animating premise of this approach is that the city is at once the largest unit capable of initially addressing the many urban, architectural, social, economic, political, natural resource and environmental imbalances besetting the modern world – and the smallest scale at which such problems can be meaningfully resolved in an integrated, holistic and sustainable fashion. As that human settlement possessed of the minimum density supportive of urbanity and organized public life and as the potential nodal point of the many countercurrents convergent with an ecological worldview, the city is the locus of sociality, local economic production and exchange, responsive architectural design and political participation – precisely the ingredients necessary to weave together the social change-oriented groups and movements for institutionalizing ecological sustainability on the lowest meaningful level and scale.

Because sustainable cities may eventually become the Archimedean point for the sustainability of larger social wholes (including the globe), a fifth principle is constituted by the idea that *all lower imbalances are to be negotiated outwards.* That is, since the sustainable city initially has to operate within larger unsustainable systems beyond its boundaries and will never be completely autarkic economically, culturally, politically or otherwise, any imbalances and instabilities overflowing its continuing efforts to maintain homeostatic balance will have to find their sources of rebalance and restabilization as elements in the social system at the next larger scales. As part of an overall strategy of ecological and social sustainability, this principle has the potential to create successively higher levels of sustainability by starting from the smallest scale and potentiating the first of the postmodern sustainable cities, which may in the process have the contagious effect of working its way through ever larger systems.

What are the implications of these alternative sustainable cities' principles for revisioning sustainable development? First, a sustainable cities strategy overcomes the tendency of globalist experiments in sustainable development to be overwhelmed or absorbed by the continuing tide of unsustainable relationships, processes and practices. That is, piecemeal reforms and innovations supportive of sustainable development scattered here and there throughout the world give little assurance that such diffuse changes will have a net-positive impact within the dynamic international political and economic system. In offering the strategy of sustainable cities as an alternative strategy and as a catalyst to long-term sustainability, we are advancing a critical point. Because the true or ideal city has historically been identified with the idea of self-sufficiency in economic needs and equilibrium of social forces (Mumford, 1961: 316), the city as a social organism presents itself as the smallest meaningful scale where the many dimensions of social and ecological sustainability can be put in balance without exporting negative externalities to other levels at all – without exporting them in a manner that cannot be absorbed as positive resources by the next higher level.

In the process of aggregating human beings into a relatively small area and providing the necessary forum for civic life, the sustainable city has the advantage of being the medium for decentralizing and localizing economic production and commerce and thus preserving the social surplus of the local economy for community self-sufficiency and self-enhancement. This argument converges with Daly and Cobb's effort to reground *homo economicus* within the network of social and natural relationships constituting the embedded individual and is the next logical step from their alternative economic model emphasizing communities within ever widening communities. It also overturns the inherently abstract notion of planetary citizenship within globalist notions of sustainable development found even in the writings of Hazel Henderson and replaces it with the idea of the person-in-a-humanly-scaled-place (i.e., the sustainable city), which allows long-range commitments to global sustainability to find a comprehensible, meaningful and delimited sustainable space and place.

In addition, a local decentralized economy organized around soft-energy path options, sustainability-oriented technology and reskilled citizens and workers establishes the basis for overcoming the historical antagonism between city and countryside, humankind and nature, economic growth and environmental health, as well as other dualisms that have rent the modern world and the industrial-technological worldview. In so doing, the ecological city in its many versions can serve as a working model whose benefits, lessons and consequences can radiate outwards, touching more and more societies around the world.

Still, to become a genuine model and stimulus to broader restructuring in society and around the globe, the sustainable city must literally find its place or places in the evolving world of contradictory possibilities and risks. If it is not to succumb to mere utopian illusions or become the victim of pseudorealist cynicism, the sustainable city must build on experiments and changes in the existing urban materials and fabric of real cities (Levine and Radmard, 1990) and be constructed out of the concrete dreams from the past, present and future by change agents who are willing to translate their hopes into social processes bent on molding the sustainable city as the true home of humankind. However ambitious and global the scope of their hopes and dreams for sustainability may be, these change agents will never lose sight of John Todd's insight that "elegance of solutions will be predicated on uniqueness of place" (cited in Orr, 2002: 28).

The strategy of sustainable cities also has the merit of bringing the momentous political tasks of struggling for global sustainability back to humanly scaled dimensions. A global politics of eco-sustainability is as amorphous as it is unthinkable. While Henderson's interpretation of the dictum "Think Globally, Act Locally" offers a promising avenue for reciprocally tying planetary and local dimensions to the program for advancing worldwide sustainable development, we would still argue that sustainable development needs a place and a space. For us, the sustainable city provides that bounded place-form (Frampton, 1983: 16–30) and serves as the locus and anchor to global networking. As Hannah Arendt has pointed out, "only where men [and women] live so close together that the potentialities for actions are always present will power remain with them and the foundation of cities, which as city states have remained paradigmatic for all Western organization, is therefore the most important prerequisite for [legitimate] power" (Arendt, 1958: 201).

As the historic civic forum for educating democratic citizens in the skills, rights and duties of public life, the city acts as the nodal point through and within which the political battles for sustainability at all other scales must be fought. Drawing on the unfulfilled dreams of the ancient Greek city-states, the shining example of the medieval Italian hilltowns and the many uncompleted projects in urban thought and literary imagination, the sustainable city presents itself as the top priority on the new agenda of the twenty-first century and beyond.

Chapter 2

THE SUSTAINABLE CITIES MANIFESTO

Manifestoes have been a vehicle used by avant-garde groups and movements in politics, architecture and other artistic communities to chart out and publicize their grounding vision. The Sustainable Cities Manifesto was the founding statement of a developing project begun in 1987 to institute a multidisciplinary Center for Sustainable Cities on the University of Kentucky campus. It was impelled by our sense that a document tracing its critical inheritance from the past and enumerating a set of key principles and supporting arguments informing the vision of sustainable cities of the future was needed.

Urban thinkers from Aristotle to Lewis Mumford and Jane Jacobs have identified population and spatial density as the hallmark of the true polis or city. Like sustainable cities, the dense or compact cities – particularly large, heavily populated metropolises – have their own origins, influences and precursors. While previous studies have indicated that sustainable cities have been and are likely to be dense cities, urban density in and of itself is not a definitive indicator of sustainability in historic cities, current cities and cities of the future. Instead, the operative principle guiding the design and governance of sustainable cities is the striving for ecological and social sustainability. Certainly, the sustainable city treats urban density as an enabling element, but its distinctive design feature lies in its commitment to the principle of homeostatic balance within the city and between urban habitat and nature. In this sense, the sustainable city subsumes the many positive elements of the dense or compact city while regrounding those elements within a more encompassing framework of ecological principles and democratic commitments.

Pre-Text

The theoretical origins of incipient models of the future sustainable city owe much to three key figures and no less to three programs for renovating urban habitats. In England, Ebenezer Howard and Patrick Geddes were important forebears of the sustainable city models. No less critical to the development of the framework informing our project is the work of Lewis Mumford.

At an early stage in the conceptualization of a theory of sustainable cities and an operational definition of sustainability, we worked to recover past efforts embodied in Howard's garden city movement, Geddes' Outlook Tower and Mumford's regional planning association program. We did so to bring urban planning into scale with the requirements of the values of true urbanity and in line with the imperatives of the ecosystem.

The critical appropriation of this urban-ecological tradition dovetailed with efforts to range over the history of the occidental city for earlier models of urbanity uniting concerns with human settlements, reflecting urban values and human habitats and achieving harmony with a more encompassing set of relationships with nature. From this historical survey, the medieval Italian hilltown – acknowledged by Max Weber as the historic exemplar of true urbanity (Weber, 1958: 54; and Schwartz, 1985: 545) and likewise considered by Mumford as the momentary and historic "manifestation of the triumph of 'communitas' over 'dominium'" (1961: 252) – captured our interest and imagination as the potential model for grounding key aspects of the future sustainable city.

Like no other historical model, save perhaps the Greek polis, the medieval *comune* and its contemporary remnants continue to beckon ecologically minded urban designers. As Murray Bookchin (1974: 98–9) observed, the medieval free city of the early Middle Ages summoned the "spirit of spontaneous design" and "human scale" from the social relations and natural contours of the land to generate a continuous fabric of urban beauty and charm. In some respects, the first true compact city was the original medieval hilltown and much more. Its rich and dense urban fabric, though seemingly without coherence, belied the surface chaos of its form and assemblage when measured against the top-down planning and rationalistically conceived and imposed forms of later periods. Evolving from the day-to-day confrontations of a political process (formed by diverse and contending social groupings lacking a political center), the medieval *comune* combined an organic architecture with a social and political heterogeneity whose only shared principle among its contestants was that no single locus of power – whether nobility, the church, the guild or class – should achieve a hegemonic organizing power over the whole of the city (Levine, 1987; and Levine and Yanarella, 1991).

Even before the formulation of the manifesto, our work identified the medieval hilltown as a model and precursor of the sustainable city. "The Future Medieval City" (Levine, 1987) identified its defining features – high population density, a humanly scaled architecture, social heterogeneity, an urban-rural balance, primary political loyalties to the city as a whole, aesthetic richness and diversity of design and social commitment to durability and repair – as critical components of the modern sustainable city. This critical insight led to a series of architectural design studios conducted both in the United States and Italy focused on the fashioning of physical design models of the "city-as-a-hill." Here, the medieval city-as-a-hill transformed into a modern/postmodern sustainable city-as-a-hill, synthesizing design features from the medieval hilltown with ecological principles promoting social and ecological sustainability. Rejecting the antimodern impulse to return to the premodern context of the medieval *comune*, these design projects sought to incorporate the latest and most modern advances in materials, construction and technology to refound elements of this historical model upon modern foundations and even postmodern gestures.

A subsequent development involved supplanting studio design work using scaled-down physical models with computer-generated models, using state-of-the-art computer modeling software. With the aid of the Architectural-Engineering System (AES) program designed by Skidmore, Owings and Merrill, a research program for realizing the full potential of the computer for holistic design of families of sustainable cities was at hand. Throughout this synthesizing research program, a commitment to democratic governance and to the idea of sustainability as a process began to take shape and inform this foundational work.

In tandem with physical and computer model building, we worked to articulate a manifesto as a founding statement to ground an intellectual center, animated by its vision. Its finalization marked the conclusion of one chapter – and the beginning of another – in the long-term program for building the sustainable city of the future. That manifesto follows.

Sustainable City Manifesto: Preamble

The crisis of the modern age appears all around us. It not only reverberates throughout our formal social institutions, it also haunts our daily consciousness and everyday lives, instilling in us a sense of historical dislocation, social malaise and personal disorientation, as well as an anticipation of imminent breakdown and sweeping change – perhaps for better, more likely for worse. In this age of Promethean dreams and apocalyptic fears, fatalism vies with utopianism for the homeless minds and hearts of us moderns, convincing us that anything is possible, but nothing can be done. Large threats to human life and the environment erupt on the horizon, even as the industrial-technological engines of material plenty and social progress promise a better life for all who are able to get on board. The social infrastructure of urban agglomerations, once called cities, deteriorates under the crushing weight of drugs, violence and prejudice, while the new face of their downtowns reflect in the latest skyscrapers the polish and gleam of multinational finance and commerce. As Gramsci stated some years ago, the crisis "consists in the fact that the old is dying and the new cannot be born, and in the interregnum a whole assortment of morbid symptoms appears" (1971: 276).

Within academia, serious thought about the condition of our ecosystem and the plight of our cities has succumbed to the very habits of thinking and practice that have brought us to this alarming impasse. Postmodern architecture and late-capitalist urban planning are part of those morbid symptoms. So, too, are trends in sociology, political science, geography and economics that borrow upon the technological optimism and political cynicism of the age. Only a radical break with modernity can overcome the multidimensional crisis characterizing our epoch and hasten the necessary restructuring of the built environment of our cities and its relation to the wider natural ecology. This manifesto outlines a vision and charts a path for recuperating the unfulfilled dreams and uncompleted projects from the past and for liberating the future from the cultural compulsions and institutional imperatives of the present in the movement beyond the crisis of industrial society and late modernity.

Principles

The multiple crises of modern society constitute a global crisis

Convergent critical analyses of modern industrial society point to the multileveled character of the crisis of the modern age. Due to the rapid proliferation of modern technology and the pursuit of Western-style economic development strategies by advanced industrial societies and most less-developed nations, this multiple crisis has reached global proportions. The underlying causes of this crisis of modernity and its technocratic politics and increasingly integrated global political economy, however, lie deep in the revolutions of science, religion and economics in the early modern age, which helped to lay the foundations of the modern worldview in Western thought and the alienating and exploitative cultural and socioeconomic systems of capitalism and state socialism. In the face of mounting social and environmental evidence of the multiple and overdetermined nature of the crisis of modern civilization, a profusion of remedies has been offered to solve or relieve these critical circumstances. Warring against the higher wisdom of a new ecological worldview, some have recycled outworn assumptions of modernity in new garb and proposed a combination of technology and scientific administration as the prescription for our collective ills.

Scientism and technocracy cannot solve the crisis of modernity

This type of mindset emanating from two expressions of the modern worldview – scientism and technocracy – promotes analytic reductionism and policy incrementalism as responses to this global crisis. While these responses have attained unparalleled knowledge and control over small parts of the society and the environment, they have contributed to the greater disequilibrium of larger social relations and natural processes as byproducts of their specialized, instrumental approaches. Coupled with the pursuit of continued material "progress" stemming from a no-limits-to-growth mentality, these orientations lose sight of the ecological fact that, beyond a certain threshold, the treadmill of production (Schnaiberg, 1980) characterizing all growth-oriented social systems disrupts the Earth's balance-seeking mechanisms by wastefully drawing down nature's depletable resource stocks. It also spews forth into the ecosystem the negative externalities of industrial production incapable of being recycled into Earth's open and self-balancing system. In the face of ecological scarcity and the possibility of ecological overreach and collapse, this distinctly antipolitical outlook can only recommend a strategy of policy reforms geared to the quick technological fix – i.e., to modest palliatives emphasizing technological means for trying to solve genuinely social, political and ultimately cultural problems. This strategy postpones, but does not evade, the inevitable day of reckoning with inappropriate values, wasteful and unsustainable lifestyles and antiquated institutions.

The working hypothesis that Earth is Gaia relocates the guiding source of our life-sustaining environment from human (or divine) agency to the ecosystem

Whatever its origins, the wonder of life on Earth appears to be that, despite significant changes in global conditions and even enormous natural disruptions, the concatenation of circumstances supporting organisms and other living processes has remained essentially self-balancing for millions of years. According to the Gaia hypothesis, Earth is a self-regulating organic system striving toward a steady state favorable to the maintenance of life and is, therefore, capable of responding to new and changing needs for human sustenance and requirements for the good life. In seeking homeostatic balance, this complex living organism can adjust within certain bounds to large-scale human interventions and technological foul ups. Yet the high-technology/energy-intensive/hard-path avenues of organizing human settlements are rapidly overtaking the capacity of Gaia to cope with spiraling interventions and assaults from these sources. For it is the nature of such options to produce negative externalities, which create large, complex and increasingly portentous social disruptions and natural imbalances that are less and less capable of correction or amelioration through policy reforms or countertechnologies.

A holistic ecological vision and decentralized political economy provide the framework for restoring balance to Earth's functioning

Overcoming cataclysmic potential of those forces and orientations propelling the world toward nuclear ruin or ecological catastrophe will require proliferation of a new ecological worldview and political economy based upon ecological sustainability and geared to localism and human scale. In tandem, those countermovements in the world, working to institute such a cultural and economic grounding, must be mobilized and coordinated. In contrast to the alienating, fragmenting and reductionist tendencies of scientism and technocracy, a new ecological framework seeks to work from lower and simpler levels of interaction and disequilibria to higher and more complex levels of balance and order; to restore intentional ties along larger wholes and interwoven processes severed by Western thought and socioeconomic practices; and to set in motion policy actions that overcome energy-environmental-economic tradeoffs presently institutionalized into our political economy.

Such ecologically inspired foundations will be borne of the twin insight that the achievement of sustained ecological balance requires a holistic understanding of the crisis – if a healthy equilibrium is to be restored to parts and wholes – and must occur at the lowest social and institutional level at which it can be effectively addressed, if political thought and action are not to be overwhelmed by powerful industrial and social tendencies promoting unsustainability at the level of social and ecological wholes.

The nodal point of many promising developments for productive change is the city

The basic unit of critical analysis guiding this strategy of ecological rebalance, economic restructuring and political governance is the city. As that human settlement possessed of the minimum density supportive of urbanity and organized public life and as the potential nodal point of the many countercurrents convergent with an ecological worldview, the city is the locus of sociality, local economic production and exchange, responsive architectural design and political participation – precisely the ingredients necessary to weave together the social change-oriented movements for institutionalizing ecological sustainability on the lowest meaningful level and scale. Moreover, because the true or ideal city has historically been identified with the idea of self-sufficiency in economic need and equilibrium among social forces, the city as a social organism presents itself again as the largest unit capable of addressing the many urban architectural, social, economic, political, natural resource and environmental imbalances besetting the modern world.

At the same time, the city presents itself as the smallest scale at which such problems can be meaningfully resolved in an integrated and holistic fashion.

The sustainable city is the political Archimedean point for fostering change

In the process of aggregating human beings into a relatively small area and providing the necessary forum for a civic life, the sustainable city serves as the medium for decentralizing and localizing economic production and commerce and thus preserving the social surplus of the local economy for the community's self-sufficiency and self-enhancement. Likewise, a local, decentralized economy organized around soft energy path options, appropriate technology and reskilled citizens and workers establishes the basis for overcoming the historical antagonisms between city and countryside, humankind and nature, and economic growth and environmental health (Bookchin, 1974; Williams, 1975). In so doing, the ecological city can serve as a working model whose benefits, lessons and consequences can radiate outward, touching more and more features of modern society. By making peace with nature and acknowledging that Gaia (not the global political economy) is the more encompassing system, the sustainable city becomes the intersecting point of the many strands weaving socioeconomic equilibrium into the loom of nature's dynamic balancing processes.

The sustainable city must find a home in the world

Rejecting the notion that the city is the seat of corruption and evil (as imagined in some versions of pastoral mythology), this manifesto also rebuffs the idea that the contemporary city can become a wholly artificial construction (integral to the technocratic myth). It demystifies these myths by recognizing that the prevailing imbalances of today's cities can only be properly overcome by understanding the city as a socially constructed phenomenon within the larger and more encompassing natural processes of the ecosystem. To become a genuine model and catalyst to broader restructuring in society and around the globe, the sustainable city must find its literal place in the evolving world of contradictory possibilities and risks. Thus, if it is not to succumb to mere utopian illusions or to pseudorealist cynicism, the sustainable city must build upon experiments and changes in the existing urban materials and fabric of a real city and be constructed out of the concrete dreams of financiers, visionary city planners, dissatisfied city dwellers and other change agents who are willing to translate their hopes into social processes bent on forging the sustainable city as the true home of humankind.

The sustainable city requires special political, architectural, and social forms

Politically, the governance of the local economy by its municipal charter takes its inspiration from the older notion of the corporation as a conferral of certain sovereign rights to special persons or associations by the community charter for specific public purposes and common ends under restricted conditions. Architecturally, the physical landscape of the sustainable city will find its cues in older models of urbanity found in the Greek city-state and the medieval Italian hilltown. Socially, the sustainable city will find its means for confronting race, class, ethnic and gender divisions in the shifting balance of dispersed power among competing groups and classes and in the common meeting grounds and shared byways crisscrossing the interconnected and overlapping spheres and institutions of city life (government, commerce, education and religion).

Many traditions inspire the shape and form of future sustainable cities

The intellectual grounding and political agenda for envisaging and constructing future sustainable cities lie in the cultural surplus of the most varied intellectual dreams and utopian projects embedded, for example, in the design of Athens by Theseus and Solon; the utopia of Thomas More; the New Harmony of Charles Fourier; the decentralized and nonviolent anarchism of Petr Kropotkin; the pastoral utopia of William Morris; the garden cities tradition of Frederick Law Olmsted, Patrick Geddes, Ebenezer Howard and Raymond Unwin; the regionalism of Lewis Mumford; the feminist/communitarian architecture of Alice Constance Austin and Dolores Hayden; the social ecology of Murray Bookchin; and the eco-feminist utopias of Ursula LeGuin, Marge Piercy and Joanna Russ.

The theory and practice of sustainable city-building must have an intellectual center

The theoretical-practical stimulus of the sustainable city will be grounded in a Center for Sustainable Cities concentrating and coordinating the many interwoven tasks of thinking and doing, designing and implementing, teaching and learning, revisioning and rebuilding. In its theoretical and visionary activities, the center will serve as a forum for promoting an urban locus for ecological and social sustainability that is open to the many pluralistic tendencies in the academy and the wider society – borne of the wisdom of the premodern and antimodern, the daring and iconoclasm of the modern and modernist and the historicity and holism of the postmodern and postmodernist (in its most productive versions). In its practical and engaged activities, it will work with existing tendencies, social forces and historically grounded possibilities that afford the center and its staff opportunities to advance its agenda on projects large and small.

The center must be a genuine interdisciplinary institute animated by a common ecological vision

The center will be organized around a common vision of the relationship between the city and the ecosystem to which different disciplinary specialties provide distinctive points of access. Recognizing the city as a place in the world, the center's staff of experts will use specialized information from respective disciplines; but, in processing this information, these specialists will seek to translate diverse knowledge into holistic designs of the sustainable city. That is, they will work to synthesize their unique and fruitful perspectives on a shared reality and field of open and unfolding possibilities into compatible models of sustainable cities, to be tested against one another and in relation to the many locations where the sustainable city might find its home. Thus, the sustainable city will not seek one utopian, perfect and therefore unrealizable form, but will be generated out of many families of computer models reflecting the diversity of forms it might take and the flexible standards that will inform its designs. Overall, the benefits accruing from these studies and exercises will be sound theory and relevant policy built upon a solid empirical and multidisciplinary foundation whose dividends will be more than the sum of its parts.

The center will be guided by an interactive process for designing the sustainable city of the future

The formulation of many parallel urban design models – informed by other supportive economic, social, political and other models – will emulate in microcosm the processes of urban growth and governance characterizing the appearance of the hilltown of medieval Italy. There, historically, such cities responded by trial and error to topography, tradition, local need and emerging desire. In this historical process, poor decisions were modified and good decisions institutionalized, virtually guaranteeing the emergence of robust cities supportive of a public life and a local culture that remains a latent ideal and a haunting presence in urban history. Guided by an interactive design process involving citizens as well as specialists, emergent urban models will be designed with the principles, tools and rules of ecological sustainability at the forefront of model building and experimentation. With the unfolding of the design process different models will be created at different scales, from construction details, dwellings, groupings, public and private space to neighborhoods and then to the entire city as an object and organism. Aided by advanced computer modeling programs and powerful memory capacities, these many and widely varying models will be stored and intermeshed so that the formal, social and ecological effects of any given intervention can be traced throughout any single sustainable city model. The interactive design process permitted by advancements in computer modeling technology will enrich and enlarge our appreciation of the many facets of the sustainable city as a social process, an architectural design problem, an economic enterprise, a feminist issue, a political task and an academic challenge.

The center's ecological vista and multidisciplinary agenda compel an enlargement of professional identity among its staff members to encompass their role as knower and actor

The fate of academic imagination and intellectual creativity bearing on the problems of everyday life in the modern world has not been a happy one. Even the twentieth century's best architects – who have dreamed large dreams and drawn bold designs – have invariably had to content themselves with the consolations of high professional esteem and a measure of intellectual immortality. Even the most audacious urban planners who have formulated theoretically coherent and structurally sound models of new towns and renewed cities have seen their designs distorted through the tough medium of urban politics. They have seen them turned into bastardized urban constructions supportive of the very conditions destroying genuine urban form and its values of civility, balance, proportion and place. Animated by the ecological vision and research and policy agenda of the center, the research staff will seek a new meaning of "professional role and identity" that will overflow the conventional roles of technician and adviser. Embracing a new relationship between theory and practice – between knowledge and action – the center's researchers will strive to restore the lost unity of the professional scholar and academic researcher as student and teacher, as scientist and democratic citizen, as knower and actor. Hitherto the key shapers of the modern world and its supportive philosophical ideas and imperatives have called upon us merely to understand and to subserve them. The life-sustaining ecosystem and its damaged body politic beckon us to reenvision modernity and to change it!

Post-Text

With its completion, the Sustainable Cities Manifesto laid down a set of principles animating our program to revision the city as the true home of humankind in harmony with nature and to crystallize its basic ideas and unfolding agenda. Its reception generated a critical mass of intellectual expertise supportive of its urban-ecological vision and prompted a whole new set of inquiries requiring further thought and reflection. Among the many questions provoked by the manifesto were the following: What is the operative meaning of sustainability underpinning the manifesto? How can the design and implementation of sustainable cities confront the problem of existing urban agglomerations populating the globe? How are we to avoid the more authoritarian-technocratic tendencies of policy planning and computer modeling in any strategy of sustainable cities? What is the relationship between the worldwide movement for sustainable development and strategies for sustainable cities? How is the quest for sustainable cities to overcome the pitfalls of utopian thinking?

One of the most vexing problems posed by the larger dialogue generated by the manifesto was the meaning of ecological sustainability underpinning sustainable cities and its implications for the design and functioning of such human settlements. Our emphasis in the manifesto on homeostatic balance within cities and between the city, the ecosystem and the design of ecological balances at the smallest appropriate scale (i.e., at the scale of the city) proved suggestive, but not satisfactory, to kindred sustainability scholars. As we searched for more adequate grounds for the sustainability component of our urban vision, our review of ecological literature led us to the work of Eugene Odum (e.g., Odum, 1969, 1989) and others (especially Redclift, 1987) who had explored natural ecosystems and had developed the ideas of "climax systems" and stages of ecological succession to help define the nature of ecological sustainability.

In the context of elaborating more fully the five operating principles or regulative ideas underpinning sustainable cities (Yanarella and Levine, 1992a), we drew upon these ecological writings and embraced a definition of ecological sustainability. This definition saw it as a condition where complex ecosystems (such as the tropical rainforests) achieve ecological sustainability, homeostatic balance or what ecologists call "climax systems" of high diversity, large biomass and high stability through protection from rapid change and "through shifts of energy flows away from production and toward the maintenance of the system itself." By contrast, we argued, "human settlements typically seek to stall such ecosystems in early stages of ecological succession, where the yield of products is high but where the

Table 2.1. Operational definition of sustainability

Sustainability is a local…	Sustainability needs a place to happen. Although problems aggregate and become manifest on a global scale (e.g., ozone depletion, global climate change), offenses to the environment are produced locally. When dealt with locally – where "local" means the city/region – the neutralization or reuse of all negative by-products must be considered part of the price of doing business. The earlier history of our civilization is the history of city/regions – largely autonomous towns that gained virtually all of their material needs from their local countryside and had to maintain the quality of the countryside in order to sustain their way of life. From this perspective, sustainability can only happen at the scale of the city/region – the largest scale capable of addressing the many urban architectural, social, economic, political and other imbalances besetting the modern world and simultaneously, the smallest scale being that at which such problems can be meaningfully resolved in an integrated and holistic fashion.
…informed…	In order to be able to maintain the quality and the productivity of the local region and its countryside, one must understand the consequences of the metabolic activities occurring within the city/region. Earlier towns operating within a largely closed system received rather rapid feedback as to the consequences of their activities. Because almost all activities manifested locally, causes and effects related to those activities were quickly understood. When imbalances threatened the city/system, they were noted and adjusted locally. In the modern world there are effectively no local boundaries and positive activities at a small scale may well have negative consequences at larger scales. By using modern means, however, we gain powerful tools both to design and monitor major energy and material flows and to model the projected implications of different processes we might choose to include in our city/region.

...participatory...	Sustainability is a process by which a local community can decide how it will afford to live within its natural budget and the limits of its own creativity. Such a process starts with the principle that sustainability is nonnegotiable, where in principle everything else is negotiable. That means that all participants in the process must agree that the health, equity and viability of the city/system are the preconditions for any other decision. As the sustainability process proceeds, stakeholders increasingly realize that they share a common destiny and that significant synergies will result from their creative encounters and negotiations. Through much iteration, the city/region becomes understood more as an urban ecosystem and less as an adversarial, zero-sum game. Eventually, the players become partners and more focused on building common wealth.
...balance-seeking process...	The problem with our existing economic system is that it has no built-in mechanism to insure its own long-term survival. It is not designed to pursue balance. As noted above, natural ecosystems in early stages of succession are also designed to maximize production at low levels of diversity, but as such systems mature, and organic material accumulates, the emphasis shifts away from production and toward maximizing diversity, resilience and maintaining internal balances. This needs to be the model for human ecosystems.
...operating within a Sustainable Area Budget (SAB)...	In the past, nature was assumed to be so vast as to be able to comfortably absorb any and all offenses that humankind's activities dumped onto it. It is now clear that we have long since exceeded many of nature's capacities. The Sustainable Area Budget is our concept for the natural budget in land area available for each city/region to support its way of life. A simple determination of the SAB for a city/region goes something like this: simply divide a country's total land area by its population and multiply by the number of people in the city/region.
...exporting no problems beyond its territory or into the future.	The key idea here is that when the prior part of this definition is realized, such a city/region will effectively export no problems beyond its territory or into the future. On the other hand, even this circumstance is negotiable given our fifth operating principle for sustainable cities, which states that "imbalances are to be negotiated outward." This means that in some cases an imbalance may be exported from the city/region, but only if its rebalancing can be accounted for by an agency beyond the scale of the city/region.

Source: Dumreicher, Levine, Yanarella and Radmard, 2000

stabilizing elements of organic matter and biomass fail to accumulate" (see page 7 of this text). "High production within these ecosystems, then, comes "at the cost of confounding nature's strategy of maximum protection or adaptation" (Redclift, 1987: 18). On this view, we claimed, the modern unsustainable city could be seen as suffering from arrested development at an immature stage where growth and change are favored over stability and sustainability. By implication, the sustainable city of the future would have to mirror more closely the climax stage of mature ecosystems in its urban-architectural form, its socioeconomic form and other elements.

In this essay on the sustainable city's operating principles, we also sought to advance our understanding of sustainability beyond the notion of homeostatic balance in two other ways. If we may define this balance-seeking principle in terms of the argument that a desired activity or process can take place in a larger system only by finding its balances within that more encompassing system, we argued, first of all, that such a principle must be supplemented by the principle of chaotic excess or extravagance – that is, the idea that a natural process or social system cannot avert the action or intervention of those forces of chaos and disorder because they are always already there. Homeostatic balance within ecosystems implies not stasis, but dynamic equilibrium; and for us, it is the existence of contradictory natural forces and destabilizing social processes that are constantly threatening a moving balance and order within systems and that injects a salutary tension into any process striving toward equilibrium and stability. These underlying elements of chaos, as Prigogine and Stengers (1984) claim, act as the great creative forces of the natural and social world. The other revision on the conventional interpretation of homeostatic balance developed in "Five Operating Principles of Sustainable Cities" is the idea that in larger social wholes (including the globe) all lower imbalances are to be negotiated upward – or, better, outward. That is, since the sustainable city will initially have to operate within large unsustainable systems beyond its boundaries and in any case could never be completely autarkic economically, culturally, politically or otherwise, any imbalances and instabilities overflowing its continuing efforts to maintain homeostatic balance will have to find their sources of rebalance and restabilization as elements in the social system at the next larger scale(s). Thus, as part of an overall strategy of ecological and social sustainability, this principle has the potential for creating successively higher levels of sustainability by starting from the smallest scale potentiating the first modern sustainable cities, which in turn can have a contagious effect of working its way up and through ever-larger systems. Finally, in our thinking, research and design work, we came to realize that the city as the proper scale of sustainability in the world must encompass the region (or as we later suggest, partnerland) in order to take into account the agricultural needs of the city's inhabitants. As such, we must then reconceive the true scale of urban sustainability as the city-region.

But where does the sustainable city approach stand in relation to the ugly, formless, unsustainable urban conglomerations sprawling over the world's surface that are defacing and polluting its natural landscape? Inspired in part by Ebenezer Howard's plan (1965: 151–9) for turning late nineteenth-century London into a web of garden cities, our efforts to confront these cancerous urban "forms" led to the idea of constructing sustainable village implantations into old and dilapidated sections of today's megalopolises in order to inject healthy sustainable tissues into the body politic of these apparently terminal social creatures (Levine and Radmard, 1990).

In its earliest expression, the idea of sustainable village implantations took form in the context of the International City Design Competition's challenge to apply new ideas and innovative methods to the repair and redesign of an aging industrial city (Milwaukee, Wisconsin). Three different sustainable implantations were designed – an urban village, a neighborhood village and an agricultural village implantation – and had to satisfy the First Law of Sustainable Design, i.e., that each implant is, itself, sustainable in itself; and simultaneously increases the sustainability quotient of its larger environment by absorbing some of the problems or negative byproducts of that larger system (garbage, sewage, pollution, etc.), and transforming them into value (food, energy, raw materials, etc.) that is returned to the larger system. Thus, each implant becomes a locus for generating sustainable forces in the larger city. Moreover, by consuming such negative externalities and generating value, each implant creates surpluses of capital for the construction of additional implants that would proliferate throughout the rest of the city.

Eventually, as new implants increase the sustainability quotient of the entire city, a point would be reached where the unsustainable industrial city would pass the threshold of sustainability and become a mosaic of sustainable parts transformed into a sustainable whole. Questions of political agency (Who initiates the implantation process?), economic support (How are the startup costs financed?) and administrative control (Who guides and monitors the process and establishes its direction and pace?) would remain open and unanswered, although even in its early formulation, a commitment to an interactive, participatory design process was articulated.

From the original manifesto to our most current writings, a concern to avoid the authoritarian temptations of the modern, elitist, top-down planning ethic has remained a persistent issue. Our attraction and fealty to the spirit of governance in the medieval hilltown – characterized by its lack of a hegemonic political center, a series of partly contending, partly overlapping social groupings vying for power mitigated by a higher commitment to values and vitality of the city as a whole – have continued to influence our thinking in working towards an outline of a governance model for the design and implementation of sustainable cities of the future. This spirit is especially important since advances in computer technology open up both greater possibilities for elite domination and social control, and greater opportunities for extending democratic participation, decentralized power and decision making in these and other activities.

In recent computer modeling projects and in writings about the process of modeling and design of sustainable cities, we have striven to fashion a process that is open-ended and participatory. There (Levine and Yanarella, 1991; Levine, Radmard and Harper, 1991; and Levine, 1992), we pursued a design process that puts the city in the computer – and the computer in the city – and placed computerized technology in the hands of those most immediately affected by the plans, designs and decisions involving the evolution of the future sustainable city.

Neither the computer models of sustainable cities like the city-as-a-hill, nor the various implantations mentioned above, nor others being generated, has congealed into unchanging, reified forms. Rather, the design process of these models are seen as part of a democratic policymaking approach where such models would be assembled and disassembled many times before actual construction would begin. Any issue, any value, any proposal could be put on the table and modeled on the computer, thus creating many parallel models tracing their probable implications for future urban form, ecological and social sustainability, and quality of life.

In our view, such a process would be a more modern and civilized version of the processes of urban growth and political governance characterizing the medieval free cities. In the historic process, city growth responded to topography, history, local need and emerging desire. Over time, poor decisions were modified while good decisions were institutionalized, virtually guaranteeing the emergence of vital and livable cities that have become recognized as some of the most supportive of public life and cultural development the West has ever seen.

This design process for generating and building the future sustainable city also has also shunned another temptation of urban and regional planners of the past: utopianism. Although we have written of the sustainable city as a necessary utopia (Levine, 1989), in its more fundamental thrust and design the sustainable city is not utopian at all. In contrast to the great utopian dreams and schemes of old, the vision and program launched by the sustainable city manifesto seek to work with the prevailing contradictory forces, giving form and substance to the future as the key element in keeping the process alive and vital and getting us from here to there. It critically appropriates urban values, democratic processes and social forms from older historic models and incorporates them into design processes using the most advanced technological means for refunctioning those values, processes and forms in the unfolding and open-ended future. Its synthesizing designs seek not to achieve perfection in any of its parts, but a durable, mature, yet dynamic sustainable condition for humankind's true home, made up of component parts that are good enough to sustain the larger whole.

Finally, the sustainable cities project disclaims the utopian urge by asking those who peer honestly into the future: where are present-day unsustainable practices taking us? Is the pursuit of sustainable cities not far less utopian than the belief that planners and politicians can continue on our present unsustainable course and deal with the accumulating problems they will find in another decade or two? (On this argument, see Yanarella, 2009.) Far from being utopian, we think, the strategy of sustainable cities appears to be the only realistic way to achieve future humanly scaled and eminently livable cities that are in tune with the needs and limits of our earthly mother.

Chapter 3

VARIATIONS ON A "GREEN" THEME: OVERCOMING SEMANTICS IN THE SUSTAINABILITY DEBATE

Sustainability is an idea whose time has come. Clearly, in an age of mounting finite resource scarcities, rapid climate change and continuing global population growth – combined with the growing clamor for economic "development" Western-style – the sustainability movement is not going to go away. Sadly, the meaning of sustainability and sustainable development remains highly contested and subject to ongoing and fierce dispute. Today, this state of affairs is evidenced by the growing shift away from the language of sustainability and its variants to the increasingly popular – and easier to swallow – term, "green."

In his latest book, *Hot, Flat, and Crowded: Why We Need a Green Revolution – and How It Can Renew America* (2008), author and *New York Times* columnist Tom Friedman both criticizes this confusion over "green" vs "sustainability" and abets it (dubbing the needed sustainability transformation in book subtitle and text as a "green" revolution). His critique in most respects is right on the money. Pointing to the proliferation of books and popular magazine articles on the many ways of "going green," he scolds: "In the green revolution we're having, everyone's a winner, nobody has to give up anything…That's not a revolution. That's a party. We're having a green party" (Friedman, 2008: 251). What really separates a "green party" from a genuine sustainability revolution? And what light do these two terms shed on the other gradations on the environmentalism to sustainability continuum that have entered the global sustainability debate?

"Green" vs Sustainability: How They Differ and Why It Matters

First things first: how can we overcome the confusion between "green" and sustainability? Although we often use these terms interchangeably, these two concepts mean different things. For one thing, "green" is typically associated with individual products and processes that seek to "pick the low-hanging fruit" (see Chapter 4). This is available in seemingly abundant supply in a country like ours, where waste remains a scandal in so many realms of commerce, industry and everyday living and where, for many Americans, profligacy continues to be a proud and thoughtless feature of a dominant consumer lifestyle. Landfill sites throughout America continue to swell even though the mantra of reduce, reuse and recycle is intoned in television commercials and on elementary school blackboards. "Green" practices are ideologically safe practices that do not fundamentally disturb the driving forces of economic growth and corporate profit making upon which planned obsolescence is tacitly built. By contrast, sustainability is tied to whole systems, of which individual consumer products and other commercial materials are a part. Its imperative breaks through the ideological veil of mass production and thoughtless consumerist consumption without end, calling for a cultural change in the definition of human need and the renovation of our competitive individualist orientation to other individuals and toward possessions (Daly and Cobb, 1994).

Secondly, "going green" distinguishes itself from sustainability in that, conceptually, it balances precariously on one leg (environmental health and well-being or economic vitality) of the sustainability tripod (economic vitality, environmental health and social equity), while sustainability rests securely on all three legs of that tripod. Whether referencing the sustainability tripod or the "triple bottom line" (the other sustainability metaphor), speaking in the vocabulary of "greenness" (green-talk) allows the environmental planner or "green" activist to focus a much more narrow agenda for change while leaving in abeyance the more politically sensitive and clearly more upsetting social equity leg of the tripod (e.g., Munasinghe and Cruz, 1995: 89). Even putative sustainable cities programs being promoted in the United States and in Canadian cities – e.g., Sustainable Chattanooga (Bartling and Ferris, 1997) and Canada's Hamilton-Wentworth "Vision 2020" program (Yanarella, 1999) – pay only lip service to policies addressing questions of equity and fairness. Sustainability, at the very least, is built upon a core meaning that makes the pursuit of all three of these legs or goals necessary and compelling.

Thirdly, "green" is popular and easy – as Friedman shows – because it connotes quick, easy and inexpensive steps to make the world *less unsustainable* – i.e., by its deployment of tactics for reducing the environmental impact of human activity, agricultural and industrial production and our built environment. The enemy – unsustainability as a set of social, cultural and economic systems and practices – is never confronted directly. Instead, a hundred simple palliatives, twelve easy practices, six quick and inexpensive

Table 3.1. Green vs sustainability: A typology of differences

Dimensions	Green	Sustainable
Relation to sustainability tripod	Only one leg (environmental improvement)	All three legs (environmental health, economic vitality, social justice)
Focus	Individual components	Interplay of individual components and whole system
Tactics/strategy	Tactical application of activities that involve "picking low-hanging fruit"; promoting individual changes and reforms to make world less unsustainable	Strategic discovery of the proper scale that will make successive policy steps and action easier and less costly by designing and implementing a sustainable, self-balancing system
Political orientation	Conventional, "pragmatic realist", reformist.	Innovative, visionary, revolutionary ("going to the roots")
Scale	Individual devices, products, indicators, practices, buildings as most tractable level for greening	City-region as the level at which human and social disequilibriums and ecological insults can be dynamically rebalanced
Risks or excesses	Greenwashing	Utopian fantasizing or top-down authoritarian policy action
Definition of success	Infinite progress of incremental improvements	Reduction of ecological footprint to a city-region's fair Earthshare

Source: Yanarella, Levine and Lancaster, 2009

ideas are sent into the battlefield without hope or consideration that they will really vanquish the adversary (e.g., McDilda, 2008; Uliano, 2008; Bach and Rossner, 2008). Sustainability, on the other hand, is radical – in the proverbial sense of "going to the roots" – and implies undertaking those necessary changes in our economic, social and urban processes to achieve a dynamic, virtuous and balanced relationship with nature. As we argued earlier (see Chapter 2), these sustainable balance-seeking processes never fully attain a static end state precisely because human life and social activities are always throwing up new, destabilizing challenges, which in turn must be tackled and brought into balance with the larger system. Thus, "green" evokes small incremental improvements in social practices, modern technology and human habitats, while sustainability implies a revolution in organizing our personal and collective lives and inhabiting the planet.

Fourth, "green" (like sustainability) can take on ideological expression. When fundamentally unsustainable companies or other uncaring institutions promote "green" talk and "green" practices, they easily congeal into a manipulative and deceptive ideology known as "greenwashing" (Greer and Bruno, 1997). The underlying ideological conceit of the jingle is intentional; its purpose is to perpetuate the trends and practices that are leading the United States and the world to the ecological precipice, regardless of slick advertising claims by powerful oil companies about their "revolutionary" research to "save the planet" (e.g., Exxon Mobil, 2009) or the trumpeting of "clean coal" by the coal industry (American Coalition for Clean Coal Electricity, 2009). Naturally, there is a risk that sustainability will fall into ideological excess. If we separate sustainability from campaigns of public understanding and active political involvement, or detach sustainability from the lives of real human beings and live communities and their politics, it can become a blueprint for authoritarian top-down policy action.

A fifth difference between "green" and sustainability stems from the scale and the scope of these policies and practices. Not only does "green" typically operate on only one leg or dimension of sustainability, it overwhelmingly focuses on individual products (more environmentally friendly pesticides or fertilizers; more energy-efficient windows, refrigerators or heaters) or on single facets (e.g., less polluting automobiles achieving higher mileage), while leaving alone the larger systems within which these products or aspects are embedded. Certainly, "greener" cars, more environmentally benign pesticides and more energy-conserving windows or furnaces may be good things – or at least better than their predecessors. However, the advantage of sustainability resides in its focus and emphasis on the combination of the "green" products, the sustainability oriented processes that manufactured them and the recyclable components with which they were constructed. As William McDonough and Michael Braungart have demonstrated in their book, *Cradle to Cradle: Remaking the Way We Make Things* (2002), sustainability has the benefit over "green" products and practices in the sense that such products are integrated into the larger system of sustainable production process. Additionally, their materials or modules are not merely recycled, but "upcycled" – i.e., the natural and technical nutrients of products are converted without waste into the raw materials for other "green" products or the components for other products. In this sense, "green" products and processes are, at best, a subset of wider sustainable building, farming or manufacturing processes – but not the reverse.

The differences between "green" and sustainability are not a matter of mere semantics. Activists organizing for sustainability in the streets, around the neighborhoods or on campus must understand these differences and recognize when agencies and organizations are advancing a mildly progressive and reformist "green" agenda and when they are truly pressing for genuinely sustainable ones. They also must pay heed to the adage that "the better is sometimes the enemy of the good," and that "picking the low-hanging fruit" in the name of sustainability is often merely postponing the larger and more formidable task of confronting and revolutionizing larger and more controlling systems of energy, food production, water, transportation and construction and their interaction.

Beyond a "Green Party": The Foundations of a Sustainability Rubric

Differentiating between "green" and sustainability provides an excellent opportunity for engaging in a more expansive undertaking: locating "green" and sustainability within a wider continuum that has informed sustainability discussion, debate and policymaking across the globe. The means for performing this task involves appropriating a concept and practice that have grown popular within education and student assessment. Within the field of education, rubrics guide student assessment (Stevens and Levi, 2004; and Germaine et al., 1998). Assessment specialists design rubrics to evaluate the degree of success or effectiveness of a student in learning a particular idea, developing an individual or collaborative project or presenting a student-taught learning unit. In general, a rubric is composed of an enumeration of different levels of understanding of, or competence in, apprehending a concept, a theory or a practice. In some cases, categorizing such comprehension on a hierarchical scale may indicate understanding or competence, but it also involves assigning numbers or points to signify the possibility of improvement. In those instances, each category or level clearly describes the criteria to be met to attain the score at each level.

Sustainability's attractiveness – and now, an even more highly favored supposed synonym, "green" – is based (largely) upon the fact that its vagueness is both its greatest strength and its most evident weakness. As a result, sustainability programs often engender confusion and sagging popular support because, over the long run, understanding of its core meaning and its essential practices remains obscure. We have taken a page from educational assessment specialists: the table below represents an effort to formulate a sustainability rubric, which starts at the ground level with a set of defining criteria of environmentalism (Level 0) and then moves from smart growth (Level 1),

Table 3.2. A sustainability rubric

Level 0: Environmentalism	– Bifurcates wilderness and built environment and identifies wilderness out there as environment. – Understands the relationship between economic growth and environment as a matter of balance or repair. – Perceives economic-environmentalist rebalancing or mitigation through government regulation and bureaucratically mandated standards or market mechanisms. – Involves purchase of pollution credits.
Level 1: Smart growth	– Rejects no-growth and controlled growth urban planning, whilst seeking smart alternatives to rampant urban sprawl. – Recognizes the need to curb unfettered economic growth and commercial/residential growth. – Advances palliatives to alleviate adverse consequences of uncontrolled growth, intended to elevate overall quality of life. – Pursues adoption of smart planning through neotraditional neighborhood design, high density downtown commercial and residential construction, mixed-use, high-density corridors congenial to mass transit and walking, and PDR (purchase of development rights) agreements.
Level 2: Green products, techniques, practices, policies	– Focuses on only one of the three legs of the sustainability tripod (environmental health, economic vitality, and social equity). – Focuses on individual devices, products, indicators, practices and buildings. – Engages in practices that "pick the low-hanging fruit" – i.e., that promote individual changes and reforms to make the world less unsustainable. – Orients itself to the political realm through conventional, "pragmatic realist," and reformist policies and actions. – Defines success in terms of indefinite progress through incremental improvements.
Level 3: Weak sustainability	– Embraces the rhetoric of the minimalist definition of sustainable development offered by the Brundtland Commission report. – Identifies sustainability as a never-ending road or pathway best pursued through a list of sustainability indicators marking progress toward an ambiguous and unarticulated end-goal. – Insofar as the goal of sustainability is operationalized, it is treated in terms of the metaphor of a table with three legs (economic well-being, environmental health, and social equity). – Retains the practice in policymaking of separating the realms of economic "development" (i.e., growth) and environmental protection through practices intended to mitigate the negative consequences of the former upon the latter

Level 4: Transitional sustainability	– Promotes a whole-building approach to sustainability by recognizing performance in five key areas (sustainable site development, water savings, energy efficiency, materials selection, and indoor environmental quality) through adoption of LEED certification standards and practices. – Pursues economic growth, deemed necessary, through urban growth and planning mechanisms that seek to combine good urban design underwritten by eco-sustainability subsidies negotiated with land developers to incorporate into buildings and neighborhood projects' sustainability-oriented design, and other practices. – Negotiates revenue sharing and other policy practices within a regional framework that mitigate sprawl tendencies from surrounding towns and cities adjacent to the urban growth boundary, and teach surrounding communities the benefits of moderate sustainability tools.
Level 5: Strong sustainability	– Understands that growth (quantitative increase) is not equivalent to development (qualitative improvement). – Works from the five operating principles of sustainability. – Recognizes the basic unit and minimum scale of sustainability as the city-region. – Conceives of sustainability as a local, informed, balance-seeking process, operating within its Sustainable Area Budget, and by so doing exports no negative imbalances beyond its budgeted territory, or into the future, thus opening spaces of possibility and opportunity. – Seeks to generate local-regional sustainability policymaking, metaphorically around the model of a Sustainability Game which involves building multiple scenarios as the driving process for generating sustainable solutions to urban development, land-use, site-selection and other issues.
Level 6: Existentially realized strong sustainability	– Forges an urban regime (or controlled growth coalition) organized around a local policy agenda, embracing strong sustainability and its political requirements. – Establishes agricultural partnerland employing sustainability-guided farm practices to provide essential food to the community. – While trading for critical nonlocally producible resources and products, effectively decouples from globalizing processes that would colonize the locale, and integrate it into dependent commercial relations. – Institutionalizes a planning system that includes a stockholder-driven and collaborative multiscenario-building process that feeds into and complements representative democratic institutions (i.e., discursive democracy and representative democracy).

Source: Yanarella, Levine and Lancaster, 2009

green products, etc. (Level 2) and weak sustainability (Level 3) to moderate or transitional sustainability (Level 4), and finally to strong sustainability (Level 5) and existentially realized strong sustainability (Level 6).

The placement of *environmentalism* at zero- or ground-level (the weakest or lowest plateau) in a sustainability spectrum is intended to acknowledge its role as a forebear of sustainability, while underlining that it is not really an expression of sustainability. A number of historical and theoretical writings have detailed the origins of environmentalism in the United States and this background is presumed in the condensed highlighting of its key characteristics and presuppositions (e.g., Hays, 1969; Schnaiberg, 1980; Nash, 2001; and Bryner, 2001). That it reached its heyday in the late sixties and early seventies and became ensconced in the administrative bureaucracy of the American state is by now common knowledge. Its location within this rubric is useful and important because different simplistic and poorly founded versions of sustainability continue to impart certain environmentalist assumptions – like the presumption that the relationship between economic growth and the environment are matters of balancing one against the other or that extracting funding via government taxation to institute policies of environmental repair or reclamation is necessitated by the negative externalities of capitalist activity or economic growth.

This rubric incorporates *smart growth* as the next to last stopping point on this continuum. The fact that the late eighties and early nineties constituted the high tide of the smart growth movement (especially in elite circles) is a manifestation of the growing recognition that the plight of many American cities – and indeed *North* American cities – was tied to fiscal crises of the eighties for many locales and the consequent decline of center cities (Shaw and Utt, 2001; and Barnett et al., 2007). Smart growth – though it remains popular in some parts of the United States today – was really a dumb sustainable strategy because it was intended to graft elements of neotraditional and other forms of smart planning onto a continued, but moderated, urban or metropolitan growth strategy. The result was a formula that was intended somehow to transcend the contradictions of an economic policy that accepted indefinite pursuit of local economic development (read: growth) strategies and a planning approach that sought to overlay smart urban design, transportation and town-and-country alternatives to rampant urban sprawl and suburbanization that had contributed to the problems generated by uncontrolled local growth (i.e., geographic, social and economic). Its inability to overcome this central predicament really exposes its shortcomings and reveals its class and elitist moorings.

We enter the more familiar (but contested) linguistic terrain of sustainability with the introduction of the "green" movement and weak sustainability. Since we said so much about the meaning of "*green*" in the first section, perhaps it is enough to point out that while "green-talk" emerged well after weak sustainability and is, in some sense, its predecessor and later incarnation, it is in other respects a less well-articulated version of weak sustainability and not an improved one. This is why it is placed below weak sustainability in this rubric. *Weak sustainability* has its genesis in the Brundtland Commission report and its minimalist, but widely embraced, definition of sustainable development as "development that meets the needs of the present without compromising the ability of future generations to meet their own needs" (UN, WCED, 1987: 43). While its other major contribution to the global sustainability debate, which it helped trigger, was the operationalization of sustainability in terms of the metaphor of the three-legged table or tripod, it remains mired in a latent environmentalist assumption. This assumption separates the realms of economic "development" (i.e., growth) and environmental protection or sustainability through practices intended to ameliorate the negative consequences of the former upon the latter.

Between weak sustainability and strong sustainability, we place *transitional sustainability* as that category of sustainability most closely identified with the current rage for Leadership in Energy and Environmental Design (LEED) certification standards and practices (Yudelson, 2007; and Kibert, 2007). Its virtue is that it promotes a "whole-building" approach to sustainability by recognizing the critical importance of performance measures in five key areas. On the other hand, its virtue is at once its bane as well, in that its focus is at a scale – i.e., a single building – that cannot truly be the level at which sustainability can be realized. When it is combined with urban growth and planning mechanisms, which seek to combine good urban design underwritten by eco-sustainability subsidies negotiated by land developers, it can make impressive strides toward strong sustainability – Vancouver is the most illustrious example here. (For an overview of the Vancouver model, see Punter, 2004; and Berelowitz, 2005.) Still, it risks two temptations: either promoting a narrow agenda fixated on the building, tower or skyscraper as the terrain of surrogate sustainability or reverting to a top-down/technocratic approach whose advances can be erased quickly by dramatic changes in political outlook and urban governance at the local or municipal level.

It is really only *strong sustainability* that understands that growth (quantitative increase) is not equivalent to development (i.e., qualitative improvement) and – we would argue – that the city-region is the basic unit of sustainability and its minimal scale (see Chapters 1 and 2). Rather than retain the environmentalist assumption, which seeks a balance of economic growth and its consequences with environmental protection or repair, strong sustainability opts for a strategy where sustainability and economic development are integrated fully from the beginning. Its definition of sustainability also surpasses meliorist and tendentious definitions by understanding sustainability as a *local, informed, balance-seeking process, operating within a sustainable area budget, and by so doing exports no negative imbalances beyond its budgeted territory or into the future.* Through this process, strong sustainability opens spaces for possibility and opportunity. This implicit distinction between weak sustainability and strong sustainability departs in some respects from other ways in which most have framed these two terms. Within debates in the field of economics, the distinction between these two versions turns on the issue of the substitutability or nonsubstitutability of natural capital for other kinds of capital. (For insight into this technical dispute, see Neumeyer, 1999.) But, to our minds, it underlines two critical elements: the proper scale for sustainability work and the holistic, systemic and synergistic character of strong sustainability.

Finally, at the top of the rubric (or the other end of the sustainability spectrum) is *existentially realized strong sustainability*. The category developed here seeks to open up for exploration and strategize the relationship between strong sustainability as a condition and the political means for bringing it to realization. The presumption here is that politics is the ineluctable field of struggle and contest upon which strong sustainability will be fought out and realized – if at all. Because it is argued that the appropriate scale of sustainability is the city-region, the aspiration to bring strong sustainability to fruition in the everyday world is, and must, ultimately be bound up in the campaign to forge an urban regime or controlled-growth coalition derived from studies by Stone and Molotch and Logan (Stone, 1989; Stone, 2005; Molotch, 1976; Logan and Molotch, 2007; and Domhoff, 2005). These bodies of literature and theory have only begun to be integrated into the study of urban sustainability (see Portney, 2003) and much spadework is needed (see Yanarella and Lancaster, n.d.). The promise of this integration of sustainability theorizing and urban regime and growth coalition analysis is that it will expose the social obstacles at various levels of governance to institutionalizing sustainability. At the same time, it will illuminate the partial successes and advances of urban sustainability programs and projects around the world into a new synthesis of theorizing and practice that will mutually nourish both.

Conclusion

This chapter has sought to contribute to overcoming some of the muddled thinking and vague conceptualizing that have clouded the strenuous and even valiant efforts to bring sustainability into the world. As we have argued, only a strong sustainability framework of understanding and practice can meet the looming threats to local and global ecosystems and enrich the human and social hosts (human beings and communities) whose lives and fortunes depend upon the ecological health, robustness and well-being of those more encompassing natural processes. Behind the linguistic caviling, conceptual debates and policy controversies over "going green" and achieving sustainable cities are real and vital stakes that hang in the balance – not least of which are our lives, the lives of our progeny and that of our life-giving planet.

Chapter 4

DON'T PICK THE LOW-HANGING FRUIT?

Introduction

Sustainability and all things "green" have become the new buzzwords of the day. Noted journalist Tom Friedman has gone so far as to declare green the "new red, white, and blue" (Friedman, 2008: 170). Still, the meaning of sustainability and "greenness" remains as elusive today as it was in previous decades. Given this condition, we may ask: Could it be that some of the most hallowed presumptions and nostrums of the sustainability movement are either wrongheaded or badly flawed?

Years ago, Amory Lovins, one of the godfathers of the solar energy and energy conservation movement, issued a now-familiar dictum: "Pick the low-hanging fruit" (Lovins, 1971). As he has recently stated, we are awash with easy means of increasing energy efficiency and resource conservation, so much so that it "is mooshing up around our ankles…and the tree keeps pelting our heads with more fruit" (Charlie Rose Show transcript, 2006: 10). By this, he means that government, business and consumers should take advantage of the recently fallen and low-lying fruit by instituting efficiency and conservation measures that are most practically feasible, least costly and offer the most rapid payback. Simply put, investment in conservation – i.e., attic insulation, storm windows, energy-efficient compact-fluorescent lighting and appliances, aluminum and other recycling, etc. – provides, by far, the quickest and highest dividends to the consumers and producers across the board in reducing the load on the environment.

In this chapter, we seek to demonstrate that a genuinely sustainable strategy requires a more sophisticated set of tactics and stratagems than the mere counsel to "pick the low-hanging fruit." As we will argue, prudent economic practice and political strategy dictate that business executives and public policymakers not just "pick the low-hanging fruit" but also simultaneously begin the process of establishing long-term investment policies that embrace efficiency and conservation synergies and, more importantly, establish sustainability as an overarching process.

Sustainability as a Path that Cannot Be Traveled

"Picking low-hanging fruit" is a pathway metaphor. It grows out of the propensity of environmentalists to embrace the first tasks of a supposed road to sustainability as identifying initial barriers to sustainability and attacking them directly and individually with discrete policies (Goldsmith, 2007). Thus, if automobile pollution is seen as a major culprit of acid rain, generate fuel emissions standards to reduce those toxic gases being converted into acid precipitation. If our corporate economy and consumer society generate ever more garbage for our landfills, institute local recycling programs to reduce such levels.

The problem with treating sustainability as a pathway to a goal is that by its own logic, and on its own terms, this linear approach constructs its own set of insurmountable obstacles that make the projected end state of sustainability unreachable. Consider the various barriers that follow.

The wall of diminishing returns

Corporations and governmental bodies engaged in green programs are rapidly bumping up against the law of diminishing returns. As such incremental conservation measures emerge, the dividends reaped by these practices diminish and become more expensive and less impressive politically. For example, urban recycling programs, once embraced by citizens for their vaunted ability to prevent steep increases in landfill costs, are now approaching the limits of practical effectiveness and are prompting localities to seek more distant and expensive landfills for unrecyclable garbage, the size of which continues to escalate, albeit at a more gradual rate (Jones, 2005). In other words, postponing the inevitable day of reckoning does not mean that the day of reckoning never comes. It is just delayed, and at best we have simply bought some time. Politically, though, the answer to the question – time for what? – is usually left hanging or frittered away without being used to discover more creative, enduring solutions.

The wall of evaporating political support

The logic of taking small policy steps is extraordinarily attractive. The initial steps are real, concrete, easily initiated, not very costly, and often return large dividends. Yet, public support for such programs proves to be a fair-weather friend. With each step, policymakers must expend greater political capital to hold together voting alignments as the costs of each succeeding step grow and investment returns become more marginal. Socialized into a quick-fix mentality operating within a short-term time horizon, many key players tend to abandon their enthusiasm and commitment to a sustainable future as soon as proximate benefits confront escalating long-term costs. To illustrate, Richmond, Indiana's pioneering Sustainable Urban-Rural Enterprise (SURE) program lasted only one political cycle before its citizens tired of mounting public expenditures and voted in a mayor who terminated it. (Beatley, Brower and Knack, 1993: 16–19).

The wall of hardball or interest-group politics

One favored method of confronting environmental threats and ecological problems is governmental regulation. Typically, this approach has been used as a legislative ploy for generating meliorist policy whose oversight responsibility is then shifted to an administrative agency. As the rocky history of efforts to mandate fleet fuel economy standards in the United States has shown (Austin and Dinan, 2005), this approach bumps up against corporate lobbying pressures in the United States Congress to postpone and reduce such standards. It also encourages extended legal battles over enforcement of such regulations. So, in order to implement such standards, these standards themselves become subject to political processes involving interest group bargaining and compromise.

The wall of technological fixation

The history of the Clean Air Act and its amendment and renewal process illuminate how good intentions borne of the strategy of policy incrementalism often lead to bad consequences through techno-fix approaches. When Congress sought to reduce local ambient air pollution levels, it mandated a set of pollution level standards that allowed industry to meet those standards through least-cost methods (Ackerman and Hassler, 1981). By adopting the cheapest method available – i.e., building higher smokestacks emitting pollutants that eventually fell hundreds of miles away as acid precipitation – this technological fix solved one problem (local air pollution) while producing another (acid rain in more distant regions).

The wall of misplaced collective effort

A diffuse, incrementalist approach to environmental quality and social sustainability promotes policies that too often focus on symptoms rather than the disease itself. As against those mainstream interests that benefit from hegemonic unsustainable policies and of national and global political economies today, "Band-Aid" policies to stem that tide typically require sustainability groups to expend and mobilize enormous political energy and organizational resources to achieve highly focused ends (regulations, new agencies, etc.). Insofar as these scarce collective resources are misdirected toward the symptoms and not the disease, they are being deployed in a battle against the wrong targets.

The meliorist approach implied by the "pick the low-hanging fruit" dictum leads to the unhappiest of all states. Not only are these walls unavoidable and insurmountable, when one peers over these walls, one finds (as Gertrude Stein said of Oakland) that there is no one there. That is, even if we could overcome these barriers, we would still not find the sustainable society at the end of the rainbow. Like Jay Gatsby's green light in F. Scott Fitzgerald's novel, the condition of sustainability ever recedes as the path is pursued until eventually each sustainability resource crashes against the wall of impossibility.

Beyond the "Low-hanging" Fruit: Lovins and Hawken on Natural Capitalism

In the face of the above onslaught, no policy advocate – least of all Amory Lovins – can maintain an unqualified commitment to the "low-hanging fruit" cliché. Indeed, Lovins only retains such a hold in one facet of his work. There, Lovins continues to advance the "techno-twit" argument that continuing advances in conservation and energy-efficiency technology mean that new harvests of "low-hanging fruit" can be achieved with each new growing season of technological development. Such innovations, he says, allow companies to extract repeated sizable savings to the bottom line. To work within the metaphor, the growing technological "trees" over time bear new energy-efficiency "fruit" from Research and Development (R&D) investments that have fertilized and stimulated new growth (Shipley and Elliot, 2006).

For the most part, though, Lovins and his associates, L. Hunter Lovins and Paul Hawken, have moved on to a more sophisticated position in their recent book, *Natural Capitalism* (Hawken, Lovins and Lovins, 2000). Still committed to a pathway approach, the Lovinses and Hawken now take the position that sustainability is a multifaceted goal requiring many pathways to achieve their highest dividends through systematic integration. The metaphor they use to respond to the above criticisms is to argue that pathway analysis involves "breaking through the wall."

The keystone of their latest series of arguments lies in a belief in the coming resource productivity revolution built upon the idea of natural capital. The Lovinses and Hawken define natural capitalism as "the resources we use, both nonrenewable (oil, coal, metal ore) and renewable (forest, fisheries, grasslands)" (Hawken, Lovins and Lovins, 2000: 2). Here, they say, we might think of natural resources in terms of the natural services they provide. What is important is not "pulpwood, but forest cover, not food but topsoil; their services provide the 'income' flowing from a healthy environment: clean air and water, climate stabilization, rainfall, ocean productivity, fertile soil, watersheds, even the processing of or conversion of waste into new raw materials (e.g., the recycling of carbon dioxide into oxygen by our rainforests)" (Hawken, Lovins and Lovins, 2000: 3).

In the past, we human beings tended to think of natural capital and its services as essentially "free goods" to be used, abused and thrown away as byproducts of our industrial system. No more. The problem now is that that we are being forced by increasing resource scarcities and global warming to recognize the folly of our wastefulness stemming from the finite and nonsubstitutable character of natural capital by money, labor or technology. As a result, we must include the costs of natural capital and its depletion in our economic balance sheets by taking seriously the stewardship of natural capital. It also means taking advantage of extraordinary technological developments that will increase the efficiency and productivity of our natural resources and their use in the market economy. In addition, Hawken and the Lovinses argue, public officials will need to knock out the props that have held up industrial processes, energy alternatives and design methods artificially advantaged by hidden or overt subsidies that exclude the costs of natural capital from the price equation.

The means to righting our misplaced public and private priorities involve the twofold strategy of breaking through the wall and achieving holism at the level of subsystems. Here, these eco-Calvinists propose that by linking a series of efficiency technologies to one another and doing so properly, synergistic economic benefits will be generated such that the efficiency value of the linked whole will be greater than the sum of the efficiency gains of the individual parts. Such resource efficiency entails integration and synthesis, not reductionism and analysis. Lovins points to the many individual energy efficiencies and conservation practices that were integrated into the design of the Rocky Mountain Institute (RMI) building, yielded a synergistic structure that cost less than a conventional office building, obviated purchasing and using heating and air conditioning systems, and permitted him to grow and harvest 28 banana crops in an edifice in Snowmass, Colorado, which often reaches 40 degrees Fahrenheit below zero in winter (Charlie Rose Show transcript, 2006)!

As for holism, the Lovinses and Hawken show how the resource productivity problem of many sectors of our economy has been addressed since the two energy crises of the seventies. They point out that United States decision makers routinely tried to solve the resource, energy and environmental problems of our present industrial processes by either starting at the wrong end (e.g., putting scrubbers on coal-fired plants) or trying to squeeze out energy and other efficiencies on the margins (i.e., making cars lighter, smaller, adding catalytic converters). What is needed, they argue, is to return to the drawing board and undertake a total redesign of products and processes.

The best example of American business shortsightedness is the Detroit Big Three's opposition to integrated advancements in automobility. For far too long, the Detroit automakers have strenuously resisted increases in Corporate Average Fuel Economy (CAFE) standards. Insofar as they did address the problem of increasing fuel efficiency, they did so in an analytic-reductionistic way. That is, by leaving untouched the basic structure and major components of the automobile, they closed off linked innovations that could reinvent the automobile. The RMI Hypercar Center has spearheaded an entirely different approach – one "designed to capture the synergies of: ultralight construction; low-drag design; hybrid-electric drive; and, efficient accessories to achieve 3- to 5-fold improvement in fuel economy" (RMI Hypercar Center website). On the other hand, since it attains such synergies in only one of the subsystems and not the entire transportation system, it risks creating a component that will be overdriven because of its fuel economy and therefore may contribute little to the efficiencies of the system as a whole (the Jevons Paradox).

From Path to Process

The coming resource productivity revolution and its potential fourfold or greater efficiencies, prophesized by these energy conservation innovators, promise to dramatically change the way our society is organized, the way we think about natural capital, how we design consumer products and even the way we build cities. Yet, as hopeful as the Lovinses-Hawken scenario appears, its catalysts have yet to trigger the kind of revolution in automobility, industrial processes and energy savings that they so confidently predicted. Moreover, the sustainability theory underpinning this "resource productivity revolution" (Hawken, Lovins and Lovins, 1999: 11) scenario is still grounded in a set of strategies and tactics involving the capture of subsystem synergies that leaves larger systems fragmented and untied to an overriding sustainability principle or holistic process. That is the direction in which our subsequent analysis will go.

What if sustainability were not conceived as path taken but a process that must be embraced and built? Analytically, we consider the problem of sustainability as a two-level process. The first is the level of "moving toward sustainability" by working to make the world less unsustainable. This first dimension incorporates the tactics and strategies outlined above and involves a commitment to politically feasible tradeoffs, such as balancing, for example, air quality against jobs or economic growth. It is essentially a reactive, analytical approach having the potential of making the world less unsustainable, but not making the world sustainable. Our concept of sustainability is grounded in the second level of sustainability – i.e., "sustainability as balance-seeking process" (Dumreicher, Levine, Yanarella and Radmard, 2000: 288).

Let us review some of the principles underpinning this understanding of sustainability.

Sustainable development does not lead to sustainability. One of the central arguments of true sustainability is that orthodox strategies of sustainable development for achieving global sustainability are shot through with the same sort of shortcomings and contradictions enumerated above – deficiencies that can only be overcome by shifting to an alternative strategy of sustainable cities (see Chapter 2). We have tried to show that conventional strategies operate at too large a scale to tack down and institutionalize self-sustaining processes at smaller scales where sustainable development may be more realizable and the result more palpable (see Chapter 3). Additionally, globalist approaches – precisely because they are so scattered and do not concentrate political energy and resources in a place and space – hold little hope of achieving a critical mass where sustainability can become a balance-seeking process.

The locus of sustainable programs and activities is the city. The linchpin of our argument is that the proper scale and nodal point of sustainability is the city-region, conceived as the largest unit capable of addressing the many imbalances plaguing the modern world in crisis and the smallest scale at which such problems can be meaningfully resolved in an integrated fashion. As that human settlement possessed of the minimum density supportive of true urbanity and organized public life, the city-region is the locus of sociality, local economic production and exchange, responsive architectural design and political participation – precisely the ingredients necessary to weave together the social movements for institutionalizing ecological sustainability.

Sustainability is a balance-seeking process. Globalist approaches to sustainability are diffuse, scattered and produce indeterminate results in any sustainability quotient (i.e., the net balance or outcome of aggregate sustainable and unsustainable trends worldwide). Localist strategies to sustainability undertake sustainable initiatives locally without grappling with larger-scale state, national and global imbalances and tendencies. The strategy of sustainable city-regions locates the city-region as the place and nexus for striving to put into sustainable balance the many systems flowing within and through the city. This results in a definition of sustainability that stands counter to the vague, but extraordinarily popular definition offered by the Brundtland Commission report. For us, *sustainability is a local, informed, participatory, balance-seeking process, operating within a Sustainable Area Budget (SAB), exporting no problems beyond its territory or into the future.*

The balance-seeking process of sustainability requires building multiple working models of the city. In coming to grips with this issue, we have moved in two complementary directions: towards centralization and towards democratic participation. That is, we have looked to synthesizing processes using computer modeling to allow us to fabricate systems models of real cities so that hypothetical and synthetic sustainable designs can be generated, and their ramifications for its various facets – commercial development, housing, energy demands, etc. – illuminated (Levine, Yanarella, Radmard and Harper, 1991). The technocratic hazards of such an approach are neutralized by the democratic principle that such synthetic design and planning processes may be opened to the widest possible stakeholders to test out, bargain and fashion sustainable models whose only prerequisite is a nonnegotiable commitment to sustainability of the whole city as a dynamic, balance-seeking process.

Sustainability, then, is conceived less as an endpoint or future condition than as an ongoing balance-seeking process. For didactic purposes, we have conceived of this interactive process modeled after a game – specifically, a Sustinable City Game – where experts and citizens explore a variety of scenarios in a democratic process ultimately leading to sustainable outcomes that balance competing interests and synthesize the best aspects of differing design-solutions (Yanarella and Levine, 2000).

Not Picking the Low-hanging Fruit as a Political Conundrum – A Way Out

The main thrust of the foregoing argues against "picking the low-hanging fruit." Before we settle for a negotiated treaty on this adage, we would like to explore what militates against implementing a set of public policies that resists "picking the low-hanging fruit."

One serious obstacle to institutionalizing our alternative is the existence of gridlock in American politics causing policy stalemate. Fractious domestic politics compounded by international politics of terrorism and fear have shaped a political landscape in the United States that has inhibited the kind of fresh thinking and even outrageous hypotheses needed to break the policy immobilism that has settled into energy and environmental policies. Such a political climate only fosters cynicism and contributes to the lure of market-based "solutions" that often exacerbate problems flowing from the many shortcomings and blind spots rooted in market mechanisms (Ophuls, 1977).

Transforming our built environment into sustainable city-regions must also deal with the happenstance that megaprojects such as ours have been historically associated with authoritarian or monarchical figures who have the power and authority to initiate them without fear of mass resentment or backlash. Smaller, supposedly more practical, activities have been linked by mainstream political science with the "genius of democracy."

No less disconcerting is the prevalence of tight budgetary constraints upon social and infrastructural programs often tied to elite-driven, pseudopopulist opposition to higher taxes. These ploys inhibit high-level public investment that often yields much higher long-term benefits. Both the Lovins-Hawken strategy and the sustainable city-region stratagem are built upon heavy investments from the corporate and/or governmental realm. Only such investments will make a decisive difference in the assault on global warming or the transition to a postpetroleum world and a steady-state economy (Nordhaus, and Shellenberger, 2007).

No less formidable is the presence of powerful political coalitions organized around so-called free-market policy ideas and straitjacketed by ideological blinders that inhibit or prevent holistic planning grounded in sustainability from taking root and informing policy processes. The rise of the so-called Tea Party movement is a case in point. When such alliances are coupled with religious fundamentalists who support a "war on science," sustainability finds few points of meaningful access into policy discourse.

The three keys to overcoming these obstacles are: a robust understanding of sustainability; a cultural paradigm embracing postmaterial values and commensalist practices; and political coalitions or regimes built upon a consensus around strong sustainability. For all of its rhetorical popularity, the field of sustainability, as noted in this book's introduction, remains a highly contested terrain in mainstream politics. Perhaps the resolution of the quarrel over "picking the low-hanging fruit" should be a revised dictum – one that honors the need for short-term benefits and accrued political capital on the one hand, and a long-term commitment to sustainable processes on the other. Perhaps we should revise the popular cliché into a more encompassing adage: *Don't just pick the low-hanging fruit!*

Such a reformulation would counsel those with political influence and decision-making authority to pick the low-lying fruit and the fruit on the ground, not as a solution to unsustainability, but as the source of precious "capital" to power the investment in those large-scale undertakings. Such undertakings will assist in realizing sustainable balance-seeking processes in a manner that breaks through the walls of diminishing economic returns and ever more scarce political capital. If politics is the art of the possible it must contend with existing obstacles and sources of resistance that necessitate some measure of immediate benefit and voter buy-in. Simultaneously, it must seek to enlarge the realm of the possible by reaching for the only true mechanisms that can ultimately institutionalize sustainability as a balance-seeking process and reproduce the political conditions of sustainability from one generation to another.

Chapter 5

FROM THE CITY TO THE CITY-REGION: THE SUSTAINABLE AREA BUDGET, RURAL PARTNERLAND AND SUSTAINABILITY ENGINE

Global competition on the sustainability front has recently taken the form of contests among nations and cities vying for the honorific of becoming the most sustainable or "greenest" city in the world. Notable examples of these rivals include Vancouver, where its mayor Gregor Robertson has pledged to establish the foundations to make his world-class city "the" premier city on the global sustainability map. Meanwhile Freiburg, Germany, has sought to lay claim to this accolade through its decades-long experimentation with "green" community living. An example of one of its latest initiatives can be seen in a section of the city named Vauban, built as a "model sustainability district" (Fitzgerald, 2010: 1–2). In the Scandinavian countries, sustainability advocates have touted Växjö, Sweden, where a single power plant provides its electricity needs and the city relies on woodchips and other biomass waste for its fuel. In addition, the gases generated by the power plant go through a process that condenses these negative externalities into liquid before purifying the liquid for household needs and heating. Through this and other technologies and programs, the city has succeeded in lowering per-capita carbon emissions by 25 percent, leading to the lowest urban emissions level in Europe. These global achievements and ambitions point to the need to discover and publicize tools and best practices that other cities and towns across the world can emulate or modify to fuel their efforts to align urban development and sustainability.

This chapter begins by surveying two favored policy instruments – sustainability indictors and ecological (or carbon) footprint analysis – that have been embraced by many sustainable city programs in North America and Europe to advance their agendas for sustainability. It offers the Sustainable Area Budget (SAB) as an alternative means for encompassing strong sustainability in towns and cities that avoids the shortcomings of the former two popular approaches, but produces its own possibilities and potential problems. From there, it focuses on the necessity of expanding sustainable city programs to a regional scale in order to move toward greater localization and regionalization of strong sustainability. In this context, we introduce the concept of rural partnerland as a putative companion to the development of sustainable cities and sustainable urban implantations. Finally, in this chapter, we place the strategy of sustainable cities into the realm of policymaking by introducing the Sustainability Engine as a necessary component to involving citizens, planners, elected officials and other stakeholders in the planning of sustainable cities as a local, informed, participatory process.

Sustainability Indicators

The sustainability indicator approach is among the most popular and widely used methods by policymakers both in Europe and in North America, from local to national and even global levels, to guide strategies of sustainable development. Within this method, a list of fairly discrete measures – identified as key yardsticks of sustainability – is typically generated. For some measures, distant goals for achieving a desired level of reduction are often set. Then periodic reviews are established, where these sustainability indicators are calculated in order to assay how much progress has been made toward achieving these goals. Whether in Seattle, Washington, Chattanooga, Tennessee or Hamilton-Wentworth, Ontario, in North America and almost universally in Europe this preferred method has been a staple in the design of strategies for "moving toward local sustainability." Perhaps the most positive argument for governments employing sustainability indicators is that the process of selecting such indicators reflects collective values (Meadows, 1972), clarifies stated goals and desired outcomes (Roseland, 1998: 203) and assures that what gets measured and evaluated gets implemented and supported (Osborne and Gaebler, 1992). Still, there is a saying that the "better is the enemy of the good," and in some sense, better and good may be incommensurable. Sustainability indicators may be a case in point. In spite of the fact that they are often represented with precise numbers, such figures are only an indication of whether we are doing better or worse in a very relative sense. In fact, the scales upon which such numbers are located do not have and cannot have sustainability as one pole. Such indicators are excellent in comparing cities (such as Milan vs Rome) to come up with an approximation of which one is worse in a variety of different weighted categories or on an aggregated, ranked scale. Such rankings can be very useful to prod municipalities to perform "better" in particular sectors, especially when performance is particularly poor. But this exercise can be repeated only a limited number of times before further improvement comes up against various barriers and diminishing returns well short of any desired goals, which make the exercise appear increasingly futile. We believe such goals of necessity must be arbitrary ones, since there is no point on any of the separate scales or on the aggregated scale where sustainability can be said to exist. Rather, the scales are better understood as indicators of unsustainability, and in this sense, the name of the scales is a misnomer. They should really be called "Unsustainability Indicators," for they can be used only to assess the degree of unsustainability in existing city-regions. The point is that sustainability is not a quantity to be measured. It is an ongoing metabolic process. Indicators do not even tell us if such sustainability processes are actually operating and if such processes do in fact exist in the cities being assessed. Improvements on an indicator scale are then to be viewed as anecdotal and not to be considered as evidence of a viable sustainability process active at the scale of the city-region. What can be measured are second- and third-order effects, whose presence (or absence) are seen to be indicators of ill health in particular specialized sectors. They are not constructed to assess balances or, in the case of potential sustainable city-regions, complex interrelated balance-seeking systems. While they are still a very useful didactic and political tool, this is not enough.

Ecological Footprint

Appropriated Environmental Space and the methods that have been developed for assessing it (e.g., Sustainable Process Index (SPI) and material input per unit of service (MIPS)) are very useful tools in making local determinations to assist in minimizing the various loads on the environment. Their great merit is that they are extremely effective as analytical tools for assessing the environmental load for everything from a small-scale material or component to a factory or even an aggregated figure for a whole city-region. Moreover, metaphorically and visually, they are powerful, even compelling ways to represent the extent of unsustainability manifested by ongoing social, economic, energy and other processes within the built environment. However, they are a good deal less useful in fashioning a solution once the magnitude of the problem is recognized. For example, when Herbert Girardet (Joplin and Girardet, 1997) calculates the ecological footprint of London (using Wackernagel and Rees's assumptions) to be 125 times the city's land area, this calculation invariably makes a big impression upon readers. This is particularly since he shows that this figure approximates the entire arable land area of the British island. His figure tells us several things: that London's metabolic processes are appropriating far too much territory; that London's reliance on land is in competition and conflict with all the other towns and cities in Great Britain; and that London's use of resources and land area must be dramatically cut back. But beyond that, it tells us little about what could or should be done or how a strategy to respond to this assessment might be shaped. The implication seems to be that in the long run, London must cut back its use of resources and its load on the land by a factor of 125 times. This cannot be the case, since London must be entitled to claim more than its actual land area. But how much more? And how do we respond in terms of policy to this gross overuse of finite resources? Do we react to such a footprint calculation by buying into the Wuppertal strategy that urges an increase in resource efficiency by a factor of four or a factor of ten? Or do we adopt a more cavalier attitude, to wit: "Let's not worry what the exact factor may be. We are doing very badly and we know what direction we need to be going. So let's get started as soon as possible with any and all discrete actions that will reduce the footprint and figure out the precise goals later on."

Unfortunately, these footprint approaches tend to encourage these mindsets, urging citizens and policymakers alike toward embarking on a succession of incremental improvements. We would argue that it is this a very incremental, local problem-solving mindset that has us winning many small skirmishes and even mid-sized battles but losing the larger war. The stratagem of increasing resource efficiency and decreasing the environmental load or footprint – in the many different places and scales at which it occurs – is the sort of program best carried out by latter-day "eco-Calvinists" (i.e., technicians, management and efficiency experts). It is thereby unlikely to benefit from the needs and creativity of the larger stakeholder population that it ultimately seeks to serve. It is thus also likely to be the sort of process that in sacrificing equity issues for efficiency goals will transform this portentous democratic policy issue into a narrowly construed technical concern and thus allow it to fall below the threshold of an open and participatory process that sustainability demands.

Footprint approaches fall victim to other problems as well. Like sustainability indicators, the ecological footprints as aggregated areas are additions of causes, which in effect appropriate land areas for their long-term maintenance. Because of this, they cannot take into account synergistic effects between causes. They thereby discourage seeking and constructing complex networks of causes as a primary strategy for reducing the footprint area. A city-region is a complex network of relationships. From an analytical perspective, these relationships can be disaggregated and quantified, but relational information is lost in such an analysis. It is much easier to disaggregate a complex system than it is to put it back together again (i.e., the Humpty Dumpty effect). It is time to move from strategies that are aimed at incrementally lowering the various loads on the environment – making things marginally less unsustainable – to a strategy that seeks out and adopts strategies based on a full sustainability agenda and is committed to adopting complete sustainability processes from the outset.

Sustainable Area Budget

Most importantly, both the sustainability indicator and the footprint approaches suffer from their inability to establish an effective operating definition of sustainable development. Because they both show us in powerful ways approximately where we are, and because where we are is far from where we need to be, they can play a powerful educative role in spurring us into action. In addition, because they can be visually represented the magnitude of unsustainability problems can be readily seen. But other than a general message to reduce, reduce, reduce, these methods do not present us with a coherent program for grappling with unsustainability. Nor do they conceptualize within their individual frameworks the condition or conditions of sustainability. While Wuppertal's strategy is to maintain a positive outlook by presenting many uplifting success stories, their overall political strategy of promoting resource efficiency and conservation is insufficient to realize the goal of long-term sustainability. It is very difficult to steer a vehicle while at the same time trying to apply full braking force. It seems futile to try to institute a program aimed directly at reversing society's enormous momentum; it is better to seek a process that can build on that momentum and steer it to a new path. While consensus exists on the principles and attributes of the sustainable city-region, it is time to come to grips with designing methods and models for getting from here to there. Now is the time for the discussion to turn from how much we may need to lower our environmental loads to how to collectively restructure our places, our processes and our lifestyles so that we can continue to operate at a given, preferred level indefinitely (i.e., a sustainable condition understood as "equity extended into the future" – a phrase attributed to Herman Daly).

In order to begin the process of bridging the divide between the unsustainable here and the sustainable there, progress on conceptualizing and operationalizing the idea of an SAB is needed. The SAB can be thought of as the other side of the indicator/footprint coin – that is, as another more comprehensive way of looking at and addressing the problem of unsustainability. Its preliminary formulation goes like this: If each city-region were to have a budget – a land-based budget – from which to satisfy its needs now and into the future, then it would have a clearer understanding of its resources and its limitations. With a fixed-area budget the citizen-stakeholders of a city-region would in principle be free to negotiate their way of life within the limitations presented by their locale, their own culture and creativity as long as they do not export any harmful imbalances beyond their SAB or into the future. Bear in mind that such a specification is not a new condition. It is the basis on which all traditional towns and cities operated historically with the exception of those situations where greed, jealousy and ignorance sometimes caused them to infringe on the environmental space of others. In a globalized economic system, such local processes no longer obtain precedence. Unfortunately, our current system of global capitalism is by its nature fundamentally unsustainable. This is so because as a system, it reserves little role for equity or environmental issues. Moreover, although many governments and institutions are attempting to apply social and ecological brakes to the globalist juggernaut willy-nilly, it continues to gather strength and increasing capacity for deflecting these minor nuisances. It is not that the system of global capitalism is purposely designed to fail; rather, its fundamental thrust lies elsewhere. It is dependent both upon people and the environment and, in the long run, global capitalism by its own design must fail. This is because it is ultimately dependent upon people and the environment and therefore cannot exploit them indefinitely. Unless we create alternative sustainable structures soon, its collapse will pull people and the environment down with it.

The concept of the SAB is simple in principle. Discounting the likely effects of global warming, the land area of the Earth and of each region on it is fixed. Material resources are land based. The land budget of a city-region, in its simplest terms, is the same fraction of its country's total land area as its population is a fraction of the country's total population. This means that each resident of a country carries a claim to the balancing potential of his or her fraction of the country's land area divided by its population. For Austria, for example, with a population of 8,131,000 people and a land area of 83.858 sq. km, each resident would have this sort of claim to .0102 sq. km (1.02 hectares) of land. This land would not be individually held, but would be part of the claim of his or her city-region. So far it is simple enough. On the other hand, such an approach raises many issues, the most general of which is the question of whether it is worth the trouble to embark upon such an exercise. But is there really any alternative? Before entering this discussion, it is useful to consider some of the benefits that will accrue from following this approach.

If we were able to agree on such a land-based budget, whether it was on a national, a continental or a global basis, we would have a clear and relatively precise basis upon which to conduct a local sustainability process for the first time. With the assistance of an array of new tools on the shelf or on the horizon, the citizen stakeholders of a given city-region could negotiate amongst themselves how they would afford to live within the limits of their land budget and their own creativity. In this new type of governance process involving an expanded definition of the democratic process, they would consider their own traditions, land, climate, history, hydrology, agricultural traditions, land resources, culture, patterns of conviviality, urban and rural patterns, recreation, resources, etc. in the process of projecting alternative scenarios for their collective future. Naturally, they would also need expert assistance in constructing and projecting these alternative models. The current authors have projected what we have called the Sustainability Engine, a computer program using BIM (Building Information Modeling) technology with object-oriented systems dynamics programming that would house land based models of the city-region. Using this programming platform, stakeholders would, with the assistance of technicians, conduct their scenario-building exercises and hypothetical experiments. The stakeholders would try out their ideas and introduce "what if" questions to emerging models of the developing city-regions. In the process, they would negotiate solutions to local sustainability that could entertain many options and involve represented interests, but under the overarching condition that sustainability itself would be the one thing that is not nonnegotiable.

To elaborate, this experimentation would occur in the context of the Sustainable City Game, an interactive construct where competing needs and competing proposals could exist side by side within alternative scenarios, modeled and displayed on the computer. In addition to displaying the three-dimensional images and walkthroughs of the various developmental proposals, the Sustainability Engine would also house a systems dynamics model that would track the state of balance of the city-region-system and of its many subsystems. With this processing capability, the engine would provide feedback as to how a given proposal (say a proposed factory or farming operation) might possibly add to the economy, but simultaneously, because it would represent a change in the previous system, it would unbalance the energy-material flow dynamics of the overall system. There would then be a succession of attempts to rebalance the system using local, regenerative resources.

Possibilities and Problems

The core principle of the SAB can be simply stated: each human being is entitled to be sustained by their fair share of the Earth's land, water and air. Still, the specific application of the principle behind this concept is by no means obvious. Ultimately, the application of this principle will need to be the subject of international negotiations and treaties. But that step is some years away. Even specific applications of this principle will need to be the subject of a number of lines of research and debate. This discourse as it is pursued is likely to provide the basis for development of a better understanding of the nature of sustainability and its consequences. A number of questions raised by the SAB concept come to mind:

For a given city-region on a country-by-country basis:

- Should the SAB be defined and pursued as a global figure, thus providing equivalent resources to all peoples in all regions of the world, or would it be more practical at the outset to work on a country-by-country basis where local control would be more manageable?
- Is the SAB to be determined on the basis of total land area or only arable or biologically productive land area?
- Is the SAB to be determined by land category (entitlement of similar percentages of agricultural land, pastures, forests and wilderness)?
- Is wilderness to be excluded from calculations and permanently protected?
- Does land quality come into the equation?
- Is SAB to be restricted to the contiguous area of the given city-region?
- May SAB land be traded between city-regions?
- Can SABs be allowed to fluctuate with changes in population or is the SAB to be fixed at a given point in time and a given population?
- When people move from town to town, does that influence the SAB?
- Should SABs be standardized to the current point in time or some future point where a "climax population," a planned mature or maximum population for the subject city-region, is reached?
- How do trade agreements and alliances form between city-regions and how does this affect local balances and the determination of local self-sufficiency?

For different countries:

- Is the SAB to be calculated on a country basis (e.g., Austria), a continental basis (i.e., Europe) or a global basis? (In such varying cases the SAB is likely to differ over a fairly large range.)
- If it is to be on a country basis, is each country still entitled to its equitable globally based share of the Earth's air and oceans?

- Are sustainability partnerships to be permitted or encouraged between city-regions in the dense, wealthy countries of the north and the less populated, poorer, less consuming countries of the south?
- Such partnerships could greatly increase the SABs of countries like the Netherlands and the UK, but what about countries like Bangladesh?

Using the Sustainable Area Budget

While there are many long-term issues to be resolved, the SAB concept is ready for research and even implementation and such use could accelerate the emergence of sustainable city-regions. The SAB will be useful in helping to answer the question: In negotiating and designing the activities and fluxes of our city-region, what are the resources we have to work with without infringing on the environmental space of other city-regions or of future populations? As has been indicated, the definitive answer to that question will be a while in coming; nevertheless, the authors propose some simple working guidelines for responding to this question. Because Europe is more densely populated than other continents and because it would be premature to claim environmental space outside of one's own country, a conservative approach would initially establish local SABs from a country based perspective. Even this does not completely answer all the questions generated by this developing approach. However, as a useful starting point, it does serve as the impetus to some valuable experiments.

For Austria, for example, we would propose as a point of departure several complementary SABs:

- For growing crops on the land: the arable land area (1.397.000 hectares) divided by population (8.131.111) = growing SAB = .17 hectares per capita
- For grazing: the grazing land area (1.943.000 hectares) divided by the total population (8.191.111) = grazing SAB = .24 hectares per capita
- For forest resources: the forested area (3.240.000) divided by the total population (8.131.000) = forest SAB = .40 hectares per capita
- For CO_2 fixation: the total area (8.396.000) divided by the total population (8.131.111) = CO_2 SAB = 1.02 hectares per capita (this number of course is to be modified by or substituted by the photosynthetic capacity of Austrian flora and is to be augmented by Austria's equitable share of the photosynthetic capacity of the Earth's oceans)

Compare these figures for those of other countries, continents and global values as shown in Tables 5.1 and 5.2. To determine the available city-region SABs in the various sectors, multiply these per capita SABs by the population in the subject city-region.

Table 5.1. Land area and uses in selected regions

Country	Total area (1,000 ha)	Land area (1,000 ha)	Arable land (1,000 ha)	Permanent crop (1,000 ha)	Permanent pasture (1,000 ha)	Forest/ wilderness (1,000 ha)	Population	Population density (per ha)
Austria	8,386	8,273	1,397	82	1,943	3,240	8,131,111	982.85
Germany	35,698	34,927	11,879	228	5,266	10,700	82,797,408	2,370.58
Italy	30,127	29,406	8,280	2,750	4,347	6,800	57,634,327	1,959.95
Netherlands	4,084	3,392	906	35	1,032	334	15,892,237	4,685.21
United Kingdom	24,488	24,160	6,267	41	11,201	2,390	59,511,464	2,463.22
United States	936,352	915,912	176,950	2,050	239,250	295,990	275,562,673	300.86
China	959,696	932,641	124,139	11,420	400,000	130,496	1,261,832,482	1,352.97
Mauritania	102,552	102,522	488	12	39,250	4,410	2,667,859	26.02
Europe	506,731	489,490	126,886	13,616	81,070	163,743	703,729,741	1,437.68
World	13,387,019	13,048,407	1,380,239	131,527	3,426,531	4,178,224	6,080,671,215	466.01

Source: Levine, Yanarella, Dumreicher and Broyles, 2000

Table 5.2. Sustainable areas budget in selected regions (hectares/person)

Country	Growing SAB	Grazing SAB	Forest SAB	CO_2 SAB
Austria	0.17	0.24	0.40	1.02
Germany	0.14	0.06	0.13	0.42
Italy	0.14	0.08	0.12	0.51
Netherlands	0.06	0.06	0.02	0.21
United Kingdom	0.11	0.19	0.04	0.41
United States	0.64	0.87	1.07	3.32
China	0.10	0.32	0.10	0.74
Mauritania	0.18	14.71	1.65	38.43
Europe	0.18	0.12	0.23	0.70
World	0.23	0.56	0.69	2.15

Source: Levine, Yanarella, Dumreicher and Broyles, 2000

The Concept and Necessity of Rural Partnerland

The strategy of sustainable cities developed in this work is of necessity a theory or approach that operates at the scale of the city-region. In seeking to relocalize important facets of societal production and consumption to make the SAB meaningful and practicable as a sustainability metric, the pairing of city and countryside (or its surrogate) comes into play. In many cities and metropolitan areas – despite their implication in the prevailing global economy – a conscious policy effort has been made to influence the interaction between urban and surrounding rural areas in order to slow down or halt urban sprawl or to involve the city and region in an exchange of goods, materials, energy and social and cultural opportunities and benefits. In Chapter 10, we present a case study of the Westbahnhof railway site and its proposed buildover, which uses what we call a "sustainable urban implantation." There, we make numerous references to the relationship between urban implantations and rural partnerlands and the economic, social and cultural trade relations.

The partnership concept represents a first small step in the reconstruction of the global economy on a sustainable basis. The essential operating goal of any sustainability regime is to ensure that the movement of materials, energy, goods and services is activated by the larger, more comprehensive forces of supply and demand that incorporate ecology and equity as balancing influences on traditional economic forces. The "rural partnerland" concept is a novel way of implementing the first operational step at the scale of the city and region. That is, it is intended to help define and fill out the sustainable area of the city and region that operates with its budget – i.e., its SAB.

In instances where sustainable city projects are implanted within urban boundaries, our theoretical approach suggests that the sustainable city implantation be paired with rural partnerland through contractual means so that the spheres of economy, environmental equity and social equity become involved with one another in assuring that these elements are properly balanced within the confines of the combined territory and the population comprising its appropriated land budget. While this does not assure that the larger city and its surrounding land aligns with its larger SAB, this process does create a connected sustainability island or foothold within the urban landscape that serves to foreshadow or model the dynamic conditions towards which the city should move to achieve a sustainability balance-seeking regime within the entire city-region.

In subsequent stages of development of the sustainable city implantation (SCI), the city models and their components will become the framework for the integration of other systems including: mechanical, electrical, material and infrastructure systems, facilities management, information, energy and material flow models, economic activity, imports and exports (input/output) to the city and the modeling of the ecological balances within the city and between the city and its rural partnerland. This will be done on the Sustainability Engine.

The Sustainability Engine

The Sustainability Engine is the "autonomic nervous system" of the SCI. Both during the design process, and in the governance and management of the city, the Sustainability Engine houses the energy, material flow and process models that are studied and tested in order to advance the development of the city. As the city, its processes and its industries are studied, the Sustainability Engine provides frequent feedback on the ongoing state of the system and indicates the sectors where it is out of balance. Thus, the Sustainability Engine is a tool capable of modeling the state of balance of a city system – either a new city in its formative stages or an existing city or part of a city – and providing the means for assisting in the rebalancing the metabolism of the city. A current goal is to develop a prototype of the Sustainability Engine as a systems dynamics utility designed both to achieve such metabolic rebalancing and to encourage experimentation in this balance-seeking process.

An overview of the state of the art in urban planning software will show that the Sustainability Engine has a history and many predecessors. Its origins go back at least to the early eighties in the efforts of Dr William Jepson to design software to facilitate the work of the University of California at Los Angeles (UCLA) Urban Simulation Team. The UCLA program sought to enlist powerful Cray computers to develop a 3D interactive model of greater Los Angeles encompassing over four thousand square miles. Its pioneering work was the first such effort to fashion methodologies and computing tools for constructing realistic 3D models of large areas of urban-built form at the city scale. The purpose of their urban simulation program was to offer interactive navigation and impact forecasting tools for diverse planning stakeholders, allowing these end-users to explore alternative planning scenarios over both space and time. The promise of the virtual city was that it would permit "the Urban Simulation Team to include virtually everyone in the planning process, expert and layman alike. [As a result] designers, architects, developers and consultants [would be] able to identify real problems and remedy those problems long before the first hole on a new development is dug" (Jepson et al., 1996: 8).

Another step forward in advancing the design and structure was the development of "What if?" – a geographic information system (GIS) and planning software program by Richard Klosterman, former professor of planning, geography and urban studies at the University of Akron and president of Community Analysis and Planning, Inc. Company (now president of What if?, Inc.). Unlike Jepson's project, Klosterman's software lacked a 3D visualization and urban simulation capacity. It focused on impact forecasting using the Environmental Systems Research Institute's (ESRI) GIS software to undertake suitability analyses to allow users to generate suitability maps and tables reporting on the relative appropriateness of different locations for accommodating future land-use demands. His "What if?" program also allows end-users to project the demand for residential, employment-related, preservation and local land uses. Key stakeholders using this software could also project future land-use patterns by allocating the projected land-use demands to different locations based on their relative fit for these uses and on public policies for guiding future growth. By allowing planners, elected officials, business leaders and citizens to utilize these powers, such impact analyses can help to fashion public policies across a wide array of areas, including implementing farmland or open-space protection policies, adopting land use plans, establishing zoning ordinances and other growth controls, constructing new roads and extending sewer and water services.

More recent advances in achieving similar goals to that of the Sustainability Engine have been made by the Environmental Simulations Center's CommunityViz software program, which was funded in large part by the

Orton Family Foundation, and MetroQuest. CommunityViz originated as a research project at the University of Waterloo, then the University of British Columbia and is now a private company. CommunityViz is a community support decision-making program that comprises three core elements: a scenario constructor (now Scenario 360), which takes geographical and other data to model the physical layout and infrastructural conditions of the community problem; a Sitebuilder 3D component that develops a fully interactive real-time 3D environment allowing participants to visualize, modify and refine alternatives scenarios; and a policy simulator, which permits stakeholders to examine the impacts and consequences of each alternative scenario generated. It has been employed in dozens of towns and cities throughout the United States and has been updated at least two times. It subsequently jettisoned its policy simulation and forecasting component and continues as a group of extensions and plugins to ArchGIS software. It is used predominantly for urban, land-use and transportation planning, as well as natural resource management.

MetroQuest allows municipalities and planning agencies to use accumulated databases in a more visualized format and in the context of examining alternative scenarios in order to educate and communicate nonexpert stakeholders to the long-term impacts of the various policy choices they make in community and workshop settings. Not only does the public deliberative process foster a sense of common ownership over the result chosen; it also opens the formal policymaking process to policies and preferences emerging from multiple-stakeholder reviews and recommendations. Critics have suggested that its main weakness is that, like SimCity, the underlying model and built-in assumptions remain hidden and not subject to critical scrutiny, modification or replacement.

Except for the original architecture of CommunityViz (now abandoned), perhaps the leading example of software approaching the structure and demands of the envisaged Sustainability Engine is GeoSim. Located in Tel Aviv, Israel, the GeoSim Systems firm headed by Dr Victor Shanker was incorporated in 2000 and its core product, GeoSim, grew out of research by the Israeli armed forces into the use of laser technology for visualization purposes. It combines two elements. The first is a highly precise interactive 3D modeling capability at the scale of large cities and a scenario constructing capacity to model the physical layout and infrastructural conditions of the designate community problem. The second is an emergent ability to simulate policy alternatives to investigate the impacts and consequences of alternative scenarios generated by stakeholders. It has recently become increasingly focused on utilizing its ground and aerial laser-based digital data generating technologies and software to assist in advancing sustainable city and university campus projects and programs.

Until recently, the realization of the Sustainability Engine has been hobbled by limitations on available technology and software. That is, the structure and demands of the Sustainability Engine have remained beyond the reach of advancements in 3D visualization, GIS and policy simulation technologies. The experience of CommunityViz involving the dropping of its Policy Simulator points to prevailing technical constraints and illustrates such a barrier. It would be wrong to presuppose that existing limitations will continue into the future, given the pace of computer processor power and the ingenuity of software developers and proliferation of computer software expertise around the globe. Artificial intelligence capable of integrating all of the data required to measure, assess and note the changes necessary to restore unsustainable city-regional systems into equilibrium, where that equilibrium point is the SAB, is edging closer to the grasp of rapidly changing technology.

The other obstacle to bringing the Sustainability Engine to fruition has stemmed from a conceptual block flowing from failure on the part of sustainability experts to embrace a sustainability yardstick like the SAB as the *sine qua non* of both defining the system boundaries of the sustainable city-region and becoming the quantitative basis for conducting the sustainability balance-seeking process. Integral to the Sustainability Engine guided by the principle of the SAB is the notion that strong sustainability is the core element built into its design and, as such, sustainability is a nonnegotiable principle in its operation. While sustainability advocates have proffered the ecological footprint as one measure of unsustainability and sustainability indicators as metrics for assessing declining unsustainability, neither approach even attempts to establish an operational measure of sustainability. Incorporation of the SAB into the mainstream planning processes at the local or municipal levels and its widespread use in planning sustainable city-regions would take the design and implementation of sustainable city programs to the required level of sophistication and practice.

Conclusion

With a sustainability yardstick or metric provided by the SAB and the modeling powers of the Sustainability Engine, a planning toolkit is now possible that can develop sustainable city-regions and sustainable town or city implantations coupled with rural partnerland that moves sustainable urban design and planning from pathway to process. That is, planning practitioners and citizens will be able to implement sustainable urban designs that not only lower the level of unsustainability, but participate in a planning process that from the very beginning engages a strong sustainability process.

What would such a participatory process guided by our operational definition of sustainability look like? How would planning experts and citizen stakeholders interact within such a sustainability policymaking framework? What rules, norms and principles would inform such participatory planning for sustainability? The answers to these and other questions will be addressed in the following chapter as the archeology and architecture of the Sustainable City Game are elucidated.

Chapter 6

THE SUSTAINABLE CITY GAME AS A GAME AND A TOOL OF URBAN DESIGN

Introduction

A revolution is occurring amidst the humdrum activities and commonplace concerns of everyday life in the city. It is a virtually unseen revolutionary transformation whose signs only subliminally reach our consciousness when we enter our homes, travel in our automobiles or step into our workplaces. That quiet transformation is the dawning of the ethereal city, the signposts of which include the miles of cables and fiber-optic wires or cellular towers tucked inconspicuously (or not so inconspicuously) just above or below the skin of the city (Graham and Marvin, 1996). It is a gilt-edged revolution in that it is unleashing progressive and productive possibilities for enhancing community social relations, as well as regressive and destructive possibilities tending toward degrading further the foundations of communitarian relations and alienating inhabitants of the city from others, the physical environment and urban fabric.

Sweeping as it is, this telecommunications revolution is a part of a more far-reaching and thoroughgoing revolution operating at the global scale. Manuel Castells (1989: 7–33) has drawn out the implications of the role of knowledge, data and information for the international political economy and cities, positing the emergence and increasing hegemony of an "information mode of development" and the dawning of an increasingly bifurcated "information city." That is, at the global level, he sees a process of "informationalization" where knowledge and data become the growing basis for capital accumulation by major corporate and transnational corporate players in the global economy. At the scale of the city, he argues that trends fostering a segmentation of expertise, access and use of data and knowledge in this increasingly information-based economy are leading to the "dualization" of the city. Here, the privileged "information haves" (Robert Reich's "symbolic analysts") live in another world from the underprivileged "information have-nots" (the telecommunications poor and computer illiterates) (see also Reich, 1991: 177–80).

Compounding this spatial segregation of information-rich and information-poor is the increasing sway of the "space of flows" operating within this new phase of global capitalism and its erosion of the "space of places." In Castells' words, this new form of space, the space of flows, is

> characterized by simultaneous concentration and decentralization of people and activities. Under the new technological conditions the key features is the linkage of different spatial locations into a new spatial process that reintegrates the functional unity of different elements in distant locations through information technology. (Castells, 1999: 3)

One alarming consequence is that the space of flows, in its increasing hegemony over the space of places, is conspiring to drain local places of their social meaning by replacing their richness and diversity with the homogenizing forms and tendencies of the globalized local nodes of corporate – and increasingly transnational – capitalism in a global network of trade and commerce, telecommunications, finance and commercialized identities.

In so far as new information technologies are heavily implicated in the local-global economic restructuring process, their participation in the restructuring of the global political economy takes place in the context of

> the deployment of the functional logic of power-holding organizations in asymmetrical networks of exchanges which do not depend on the characteristics of any specific locale for the fulfillment of their fundamental goals. The new industrial space and the new service economy organize their operations around the dynamics of their information generating units, while connecting their different function to disparate space assigned to each task to be performed; the overall process is then reintegrated through communication systems. (Castells, 1989: 348)

This disjunction between functional flows and historically determined places, argues Castells, translates into the paradox of late modernity: i.e., that people live in places, while power rules through flows (Castells, 1989: 349).

As Castells sees it, the challenge presented to localities and grassroots movements by the new information economy and society is to find the ingredients and resources for reconstructing an alternative space of flows on the basis of the space of places (Castells, 1989: 352–3). This dual task involves rearticulating place-based social meaning and organizing mechanisms of social control of places over the functional logic of the space of flows. In the era of fast capitalism, the dominant tendencies have been just the reverse.

This chapter proposes one vehicle for taking up the challenge of the presently constituted and dynamic space of places posed by Castells' critical analysis: the Sustainable City Game as a democratic means of advancing a strategy of sustainable cities. It begins with a discussion of the Cold War origins of the internet and its inchoate community of designers with a view to illuminating some of the ingredients for the success of this operation and pointing to its ambivalent democratic potential. It then highlights the emergence of a variety of communal meeting places for inhabitants of cyberspace serving as harbingers of cybercities and virtual communities. Like the emergent telematic city, these "soft cities" communities have been presented by their designers and "netizens" as complements or alternatives to the "hard cities" and everyday communities of citizens in geophysical space. This analysis then turns to an outline of a participatory design process for generating sustainable cities grounded in the Sustainable City Game. Its alternative scenarios, rules of play and implications for the design of sustainable cities are then considered and a prototype of the sustainable city generated from the framework of the Sustainable City Game is offered as an example. The chapter closes by musing on the hard issues and questions raised by Castells regarding the rehabilitation of the space of places and the pacification and reconstruction of an alternative space of flows.

From Netville to Cybercities

The emergence of the internet is instructive of the profound issues of public policy and democratic theory we moderns are confronting in our everyday lives. In an odd and peculiar way, the origins of the internet point to the paradox of politics where promising and potentially liberating advances can come from dark and even authoritarian beginnings (Haraway, 1989). A popular and oft-told narrative of the birth of the internet (Rheingold, 1993; King et al., 1996; Hauben and Hauben, 1997) relates how the United States Defense Department's concern with the survival of defense communication in a future nuclear war prompted an alliance among the Advanced Research Projects Agency (ARPA), the RAND Corporation and university scientists to build the first incarnation of a primitive communications network that ultimately grew into a global network tying most of the world into its electronic embrace. Despite – or to a great extent, because of – its birth in the belly of the Cold War beast (the military-industrial-university complex) and its emanation from the ARPA (the high-frontier research and development arm of the United States Defense Department), it benefited from generous funding, an absence of centralized planning and direction, and a mobilization of computer intelligence that congealed into a kind of decentralized intellectual and social community among its key participants. Termed "Netville" by recent analysts (King et al., 1996), this nascent cybercommunity inhabited a "great divide" – a zone of freedom unshackled by governmental regulatory mechanisms and market controls and imperatives that promoted an environment of creative problem solving. This zone of freedom and opportunity fostered the incredible growth and expansion of the internet as a space of communication and research collaboration. The conditions supporting this (granted, elite) community were special ones and, in retrospect, fragile ones that were ultimately undermined by the agencies of commerce and industry as well as the withdrawal of ARPA from its protective role.

But for a time, leavened by vast amounts of financial support and heightened enthusiasm and expectation of unmatched technological performance, this coalition spun a web of communications exchange and creative innovation geared to the strengthening and protection of the Cold War national interest defined in terms of US technological superiority (King, et al., 1996: 11). However, tainted by the military values of Cold War politics, the communications network emerging from this military-industrial-university complex was heavily shaped by three core values. They included intellectual curiosity, where the discovery of solutions was its own reward; an informal meritocracy where status within "Netville" was grounded in peer-judged merit and performance; and presumptive egalitarianism, where one's technical expertise was the entry ticket to this cybercommunity.

Mitigating the anarchic and entropic tendencies flowing from open access to the network were the consensual articulation, negotiation and establishment of ground rules for being a citizen of "Netville," sharing its powers and privileges, and participating in its technical improvement. As King et al. (1996: 13) characterize this key control mechanism:

> Rules of play were the conditions to which all members of Netville should adhere while contributing to the development of the network and benefiting from the work of others. It was the rules of play that ensured that the technologies of the emerging Internet remained interoperable… Rules of play [then] constituted conditions for entry and continued presence, and were not negotiable. Though non-negotiable, these ground rules remained flexible enough so that the dynamism of the game of Internet design and technical improvement would not be stunted by rigid conventions and frozen standards.

Despite ARPA's withdrawal from its roles of institutional adhesive and financial underwriter of the internet and the subsequent invasion of commercial internet service providers (ISPs) and agencies of electronic commerce, the casual net surfer can still feel something of "Netville's" original animating spirit of creative anarchy and organic community lingering in the increasingly crowded and market-oriented medium of the World Wide Web. Indeed, this electronic frontier has become the home of many inventive newsgroups, chat rooms, multiuser dungeons (MUDs) and object-oriented multiuser dungeons (MOOs) that approximate the romantic ideal of cybercities and virtual communities (Rheingold, 1993; Kiesler, 1996).

Small wonder, since latter-day close students and observers of the cyberspace from urban architect William Mitchell (1995) to literary theorist Scott Bukatman (1993) have pointed to the startling similarities between patterns of the modern megalopolis and the design of the computer chip. For example, a recent advertisement by Intel for overdrive processors depicts bicycle speed racers riding the lanes of a computer chip that morphs into a cityscape. And Scott Bukatman, operating from this association of city and computer chip, has characterized Gibson's image of cyberspace as "an imploded urbanism – Blip City" (1993: 150).

On the internet, the development and proliferation of cybercities, cities of bits and digital communities promises to dissolve the geophysical city in the electronic vat of digital acid in cyberspace. Like the unfolding of the ethereal city, cybercities like Electronic Blacksburg, Virginia, Digital Amsterdam and Virtual Los Angeles appear at once exhilarating and frightening. Freed from the physical confines of geophysical space, digital cities offer the possibility of creating an electronic noplace/anyplace where "netizens" can congregate on the net and through their very disembodiment overcome all traces of those differences that so deeply divide us in the unsettling era of late modernity (Boyer, 1996).

Whether these digital communities can realize a new vision of democracy in the form of virtual polices populated by "netizens" in cyberspace, or will devolve into largely commercial electronic shopping malls and advertising boards (with only the most meager residues of a cybernetic public sphere) is both a thorny theoretical issue and a portentous political issue. The question involves the contested terrain of cyberspace, and it is a matter that we and others (Yanarella, 1998; Poster, 1995; Doheny-Farina, 1996; Luke, 1999) have already addressed. For our purposes, the question of different ways in which electronic spaces can be integrated into urban places presents an opportunity for exploring the import of those forces, which drive the ascendancy of the telematic city and the cybercity for contributing to the emergence of the sustainable city of the future. The continuing integration of technological advances of modern telecommunication into our common places and urban spaces is not at issue. Nor is the inescapable interweaving of soft cities and hard cities into complex and confusing hybrids. At issue are the particular form that these developments will take and their consequences for enlarging the public sphere and increasing the scope and reach of public participation in the continuing interaction between the space of flows and the space of places.

Design Process as a Participatory Game: The Sustainable City Game

If sustainability is to negotiate the poles of the local and global and the regional and the global, and if the democratic promise of cyberspace and computer technology is to be critically appropriated rather than ceded over to the globalizing forces shaping the space of flows, the sustainable city of our future must find a new process and a new context to mobilize the necessary resources to bring it into being. The historical touchstone of this new design process and new institutional context is the medieval free cities or hilltowns that dotted the landscape of Old Europe, especially northern Italy. There, the development of towns and cities responded to topography, history, local traditions and culture, emergent needs and incipient desire. Over time, poor decisions were modified while good decisions were institutionalized in the collective memory of the urban inhabitants, virtually assuring the emergence of unique and vibrant cities that became some of the greatest supporters of civic life and cultural richness we have ever seen. Political structures lacking a dominant elite and therefore approximating to communities without unity, a strong urban fabric embedded in the layers of history, an economy balancing an vital urban center and a bounteous farming countryside and a collective memory fostering collective genius within a healthy culture – these were some of the conditions facilitating a sustainability oriented process within every realm. Admittedly, the late modern city is far too complex, heterogeneous and hierarchically ordered and its shaping influences far too hidden, artificial and unsustainable for such a process to work today. As a space of places, the contemporary city has become all too caught up and misshapen by the space of flows, undercutting its social meaning and collective identity molded by its local history and culture.

The solution is not to abandon the protodemocratic values and the participatory impulses of Italian medieval hilltowns. Nor is it to acquiesce to the incipient global culture evolving from the new functional logic of the dynamic global networks comprising the space of flows. Rather, the solution is to refunction these premodern and even antimodern elements and resources on the late modern and postmodern foundations in a manner that takes advantage of twentieth-century materials and technological capabilities for realizing a new expression of this very old participatory design process.

On one level, this participatory design process has been translated by the Center for Sustainable Cities into a game called "Emerald City: A Roleplaying Sustainable City Game." In the context of exploring alternative policymaking processes, "Emerald City" offers three scenarios within which three players seek to promote competing interests and core values (see Appendix B). One scenario presents the classic liberal interest-group model where individual players pursue their interests and seek to maximize their core values grounded in individualist assumptions and a player strategy based on zero-sum outcomes (i.e., one player's gain is another's loss).

A second scenario draws on the conditions outlined in John Rawls' book, *A Theory of Justice* (1989), where players operate under a veil of ignorance of the actual role they are taking and the core values and interests they are seeking to advance. The purpose of this scenario is to show how this veil fosters limited cooperation and enlightened self-interest that incline all players to hedge their bets against being losers in the game – once the veil of ignorance is cast aside at the end – by trading down and compromising. The resulting policy chosen is an outcome that distributes benefits and costs across all players relatively equally. Finally, the third scenario outlines a form of gameplay where the rules of play encourage cooperation and a strategy of introducing external resources that enlarges the pie and encourages players to view the problem triggering policy debate as a provocation to non-zero-sum outcomes promoting every player's interests.

The point of "Emerald City" is to sensitize players to the influence of changing roles and rules of play on possible or likely outcomes. It is also intended to clarify the consequences of institutional settings for shaping the politics of choice. As an exercise in social learning, "Emerald City" seeks to illuminate the rules and conditions that contribute to policymaking processes that set in motion participatory dynamics enhancing cooperation and sustainable outcomes from problem contexts. These problem contexts would otherwise devolve into atomistic and fractious contests for competitive advantage and personal gain.

On another level, the Sustainable City Game can – and should – be transferred to real life democratic processes in actual urban decision making. Instead of players of a game, citizens and representatives of a group would become partners in a game directed toward confronting and solving real urban problems and controversies. As will become evident, an advantage of the real world Sustainable City Game is that each proposed solution would be assembled and disassembled and structured and destructured many times during the design process before it would actually be constructed. Here, the citizens of the future village would use an interactive construct to build up families of parallel models in many sectors and at many scales of the emerging sustainable village. The interactive construct would serve as a framework within which models of neighborhoods, industries and institutions would find a right size and an appropriate place.

In the generation of a Sustainable Village Implantation, a variety of computer-aided processes are employed, including CAD (computer-aided design), BIM (Building Information Modeling), GIS (geographical information systems) and systems dynamic modeling software. The process involves first assembling many different relatable and interchangeable modules dealing with energy, agriculture, architecture, urban design, industry, economics, construction, infrastructure, governance and social programs. From this point, the Sustainable City Game can be played by anyone who may be interested in the prospect of sustainable cities.

Initially, the game is played at a simple level with players who may be both lay people or experts in various disciplines or industries attempting to follow their ideas or their self-interests to construct city models or, more likely, partial models of activities (like manufacturing, energy production) or places (neighborhoods, schools, piazzas) that may be of interest to them within a hypothetical city. As the "game" proceeds, almost anything may be proposed, even activities normally perceived to be unecological (e.g., polluting, manufacturing activities). The game process is one of model building through negotiation, and is very much akin to the gameplay of the third scenario in "Emerald City." In the Sustainable City Game, a proposed structure, system or activity, to be viable within this process, either must negotiate local balance-seeking relationships with other activities (following the second operating principle of sustainable cities) or must find a linkage with larger-scaled systems or activities that assumes the responsibility for rebalancing any negative consequences of the local process (following the fifth operating principle of sustainable cities).

Over the course of the game, different players can assemble partial models of dwellings, neighborhoods, shopping streets, squares, schools, hotels, factories, infrastructure, parks and recreational facilities, churches, agriculture and so on. The first models that are constructed will likely be relatively simple ones. Each constructed model is stored in the database and its qualitative and quantitative characteristics – its imbalances and its characteristics of compatibility with other potential modules – are noted. Each stored model is a "free body" – that is, it is an incomplete open system having its own coherent internal structure. On the other hand, it has "loose ends" or imbalances at its periphery. If it were a complete closed system, it would be disconnected from everything else and therefore could not become part of the sustainable city. (Our present cities are composed of an architecture that is conceptually closed, but systemically open, thus giving cities the disadvantages of both!)

The process of assembling these partial models involves combining them with other potentially compatible and complementary partial models in such a way that these inputs and outputs, or loose ends at their periphery, become opportunities for connection. Through such combinations and connections, the growing city or system is brought toward balance. These larger models are each available in subsequent play as either starting points or default conditions for constructing new models. For example, they may be used for fleshing out a city by quickly taking previously constructed neighborhoods or piazzas out of the database and assembling a number of them together to create a rough model of a new implantation, ready for modification or redesign.

The Sustainable City Game then is a guided process in which the First Law of Relational Sustainability is conserved (while it is favored that activities and components be economical and efficient at their own scale, what is essential is that such components and activities, whether or not they are efficient, become part of a balance-seeking process or system at a larger scale). Thus, many games are played over time, creating modules and models that are stored in the computer and are available for later use in other games. They are stored both as architectural/urbanistic entities, but also in terms of their many nonspatial characteristics (both qualitative and quantitative) contained in the database. Like the models, the modules are stored as free bodies, that is, as organs that may be implanted (transplanted) within an organism, with a notation of all the inputs which would be necessary to sustain them. A notation of the outputs which either would be used as resources by the larger neighborhood or city or would need to be rebalanced at a larger scale within the city/organism would also be needed.

As the game continues to be played, the models become more extensive, more complex and more varied. Families of details accumulate at the smallest scales and families of whole cities emerge at the largest scale. The families of cities are complex and dynamic as they are assembled from compatible and potentially interchangeable details, modules and models on many different scales. Thus, a city model as housed in the computer is not a static three-dimensional form. Instead, it is a living organization of variable relationships which, as they are molded and modified, carry with them the system's characteristics and information that animate the rebalancing process as it develops the city's complexity and keeps it sustainable.

Several factors make a sustainability process workable. First of all, such a process starts with the cardinal "rule of play" in the Sustainable City Game – namely, that while at least in principle everything else is negotiable, sustainability is nonnegotiable. That means that all participants in the process must agree that the health, equity and viability of the city/system are the preconditions for any other decision. The nature of the Sustainable City

Game is to try to satisfy one's individual self-interest while maintaining the viability of the city/system. As an individual or an industry or a sector will be incapable of satisfying its own needs alone or in isolation, it becomes necessary to engage others, both to provide those means and to correct for the imbalances caused by the satisfaction of those needs. Using the Sustainability Engine, each stakeholder attempts to satisfy his or her needs and interests through a variety of different scenarios, each involving different strategies and different partners (Levine, 1994; 1996).

Strategies that throw the city/system out of balance are quickly eliminated or are rebalanced by introducing new ideas or processes with different attributes. Over time, more favorable strategies are built on and elaborated while less favorable strategies are set aside (although they are still stored in the Sustainability Engine, possibly to be revisited if conditions change at a later date). Developing scenarios are favored and pursued when they satisfy multiple interests, when they bring the city/system toward balance and when they hold the promise of fostering equity for all the stakeholders.

As the Sustainable City Game proceeds, the stakeholders increasingly realize that they share a common destiny and that significant synergies will result from their creative encounters, negotiations and cooperation. Over time, it becomes less a game and more a real economy and urban construct. Through much iteration, the city or region becomes understood more and more as an urban ecosystem and less and less as an adversarial, zero-sum game. Eventually, the players become partners and become more focused on building common wealth.

Yet if the city/system is close to balance, any major intervention is almost certain to throw the system out of balance. The problem is then to seek the means to bring the city/system back toward balance. Even a city/system that has been thrown far out of balance presents an opportunity for major interventions. In this game, there are no inherently bad moves. On the contrary, if the city/system should ever come exactly into balance, then, in a sense, the game would be over and such closure might actually be undesirable or at least premature. In any event, the design and management of the city/system are an ongoing, never-ending process. At some point in the process when all the stakeholders are working together in harmony and the economies and opportunities of play crystallize, it would be appropriate to build the sustainable city implantation.

This discussion reveals that there can really be no ecological activities in isolation, i.e., outside a sustainable system. By the same token, every activity within a city that operates through balance-seeking principles is, or becomes, a sustainable one. However counterintuitive, social problems should be regarded as opportunities, not liabilities, within the context of discussions over sustainability.

If the problems are sufficiently critical or pressing to trigger action, then those problems should be regarded as capital resources (i.e., resources that the solution of the problem must mobilize for its solution). When the problem has been solved, then the resources have been spent. As the problem disappears, so do the resources that they have generated for their solution.

This suggests that there is good reason to question the commonsense assumption that problems, because they seem to represent entirely negative entities, should always be solved at the earliest opportunity. Because they sometimes require the marshalling of substantial resources, it is important not to squander those scarce goods on short-term solutions to discrete problems. Instead, the aggregation of problems may be our principal vehicle for placing these many so-called discrete problems into a larger framework and then synthetically addressing the problem cluster on the most appropriate scale – i.e., the scale of the city (the preferred scale of the whole). Simultaneously, over the long term, the costs to the community of solving these problems, as we shall see, may actually lessen.

To illustrate, responding to individual policy issues like waste-water runoff or overcrowded streets or deteriorating neighborhoods individually and in isolation from one another as public clamor or political circumstances force one or another of these issues onto the political agenda may seem politically expedient or administratively prudent. On the other hand, aggregating them within a more encompassing framework is more likely to foster an ongoing process that views these problems in their connectedness with each other and to the urban system of which they are a part. Thus, as capital resources and political opportunities they may provide the foundations for instituting a sustainable balance-seeking process that persists over time and deploys those capital resources in a manner that challenges policymakers and citizens to overcome underlying structural conditions that simply reproduce these linked problems in slightly different forms over and over again.

The Sustainable City Prototype: The City-as-a-Hill

Discussion of the principles of the Sustainable City Game and the theory of sustainable cities would be abstract and empirically unconnected if an architectural or urbanistic framework complex enough and rich enough to support these principles and this theory could not be found. In fact, it is the other way round, with the model and inspiration for this project coming from the historic study of sustainable cities and the social, political, constructional, topological, architectural and urbanistic processes of their generation.

From 1994 to 1999, the Center for Sustainable Cities, in collaboration with Oikodrom in Vienna, Austria, was involved in a number of multidisciplinary social and design studies of a sustainable city implantation over the Westbahnhof railway site – the westernmost railroad gateway to Vienna, Austria. The particular model from which this project derives is the Italian medieval hilltown of the free city (*comune*) period. It is in these cities that the modern roots of democracy and participation were born. But it was also here that a complexity of architectural fabric and urban form emerged that has been admired by generations of architects and urbanists who have nonetheless not been able to reproduce a modern counterpart having a comparable level of responsive building and life-affirming civic space.

This proposal attempted to capture some of the most favorable characteristics of the medieval city together with the best possibilities accessible to us in our own time. Instead of the medieval city-on-a-hill, we proposed a modern city-as-a-hill, in such a way that the exterior fabric of the city wraps over the inner core of the city. In the current family of models, the city has ten principal levels, with every third level (levels one, four, seven and ten) being a major horizontal circulation level, working its way around the city as a ground surface, as well as through the city to its interior spaces. On the surface, building fabric is usually two to four stories tall above these circulation levels, so that typically no point need be more than one level above or below the possibility of access from a major circulation street. This means that few elevators are required in the building fabric; rather, elevators are mostly used to link major circulation levels and to link the surface of the city with major spaces in its interior. Linking all the levels are two sloping streets that start at ground level at opposite ends of the city and rise at a gentle six percent slope to arrive at the level ten center of this sustainable urban implantation. As they pass through major circulation at levels four, seven and ten, there is the opportunity to create piazzas and also to go through the hill to its inner public spaces organized around large gallerias and concourses. In addition, all the industry, the infrastructure and utilities, much of the larger institutional and commercial activity, as well as everything to do with transportation, movement, shipping, parking and support services are situated on the inside of the hill. Thus, the interior contains all of the activities whose large scale would make them disruptive of the rich and intimate scale of living patterns on the city surface, but which fit nicely into the large volumes of space available within the hill.

The intensity and concentration of the activity in the interior liberate the surface to project the same serene qualities and to foster the kind of public life still to be found in the streets and piazzas of medieval hilltowns. The surface is free to devote itself to dwellings, neighborhoods, small-scale commercial and institutional uses, as well as to the public realm, including the pedestrian promenades, piazzas, parks and playgrounds, which throughout history have animated civic activity. Every horizontal surface in this terraced cityscape is utilized for activities ranging from private terraces and gardens to public streets and parks. This urban model offers unparalleled opportunities for the making of place – inside, outside and in between, spaces above and below, spaces active or quiet, public or private and three-dimensional spaces – with energies and activities at many levels. It is an urban model for a modern livable, sustainable city, as well as a model for pumping new life into the existing unsustainable cities of our society. This is not a "problem-solving" approach. It is not that the problems of the unsustainable metropolis have been solved through this approach. Rather, through the synthesis of a new model and a new process, the problems have been avoided by creating a new urban prototype that affords neither the place nor the occasion for those problems to occur. (See chapter 10 for a rounded case study of this design project.)

Conclusion

As a looming political challenge of enormous proportions, cyberspace then is not an alternative to human habitation on the Earth, but merely another form of habitation within a more encompassing built environment enclosed within an even more encompassing set of relations in nature. Likewise, the space of flows neither resides in some ethereal realm nor is completely separated from the space of places. Its global threads always emanate from specific places and express their power in the countless nodal points through which their economic power and political impact express themselves and are felt in our everyday lives and the space of places. The gauntlet thrown down to those of us advancing an agenda for sustainable cities is to find the means to negotiate these complex, interwoven interrelationships between the space of flows and the space of places and among cyberspace, built environment and nature, as Castells has so well argued, through the reconstruction of an alternative space of flows on the basis of the space of places.

The Sustainable City Game offers much to those in academia who are striving to find critical tools and simulations that point the way to advancing beyond the zero-sum games that constitute the typical approach to urban planning and urban design governing many planning agencies at the local level. Conventional approaches to urban planning, the Sustainable City Game demonstrates, merely reproduce the underlying individualistic assumptions and winner-loser outcomes at higher levels, turning the solutions that have been designed and implemented to overcome existing problems into the ingredients for new and more daunting problems in the future. By moving beyond Scenario A (where individual self-interest governs) and Scenario B (where a veil of ignorance tempers such self-interest) to Scenario C (animated by the nonnegotiable principle of sustainability), the citizen-stakeholder as an actor is encouraged to explore the bases of the sustainable commonwealth. In so doing, the citizen will be able to think seriously about how real world processes can be informed by new governing assumptions and rules of play that break out of the current straitjacket of orthodox presuppositions and shopworn rules that continue to fuel the Unsustainable City Game of contemporary urban planning practice.

The strategy of sustainable cities underpinning the Sustainable City Game opens up a particularly inventive and fecund avenue for mobilizing local resources and local flows in particular places and across coordinated spaces to accomplish precisely this daunting task. As the nodal point of many promising developments in alternative technologies (e.g., urban architectural designs in solar energy and energy conservation), experiments in local-regional economic restructuring along ecological lines (e.g., Wackernagel and Rees's ecological footprint and our Sustainable Area Budget) and new democratic-participatory institutions (e.g., the Sustainable City Game), the sustainable city stands out as the most appropriate scale for social synthesis and platform for reconstructing the prevailing space of flows in an alternative social form and more promising ecological basis. As that human settlement possessed of the minimum density supportive of urbanity and organized civic life and as the potential Archimedean point of the many countercurrents convergent with an ecological worldview, the sustainable city is potentially that *locus* of sociality, local economic production and exchange, responsive architectural design and political participation most capable of reconfiguring the space of flows along democratic lines and in more humanly scaled forms.

The alternative democratic and authoritarian avenues opened up by the microcomputer, the internet and telecommunications should not deter grassroots movements and other social movements in their quest to tame the space of flows. Instead, that ambivalence should serve as a provocation for pressing onward to exploit the democratic possibilities of these new technologies in ways outlined above. While always giving primacy to the space of places in local-regional efforts (and even locally and regionally coordinated global networks) to restructure the space of flows, such groups should show no reluctance to integrate emergent electronic, computer and telecommunications technologies into the expansion of citizen participation. The assuming task of community revisioning and local-regional action is to build up the strength and resilience of local-regional flows of capital, trade, commerce and other exchanges that fashion a vibrant local identity, a robust local-regional urban-rural economy and an augmented communitarian civic culture and politics. These local resources will be the tools to restructure the flow of places currently spanning the globe.

In some respects, the greatest challenge facing the incorporation of the Sustainable City Game into the policy process will be to muster political resources and design political strategies for institutionalizing in the space of places the "rules of play" explicitly enumerated in the sustainable commonwealth scenario of the Sustainable City Game and tacitly assumed in the participatory design process for designing sustainable cities and sustainable village implantations. In city after city and town after town (in the United States especially) local ordinances, zoning regulations and building codes militate against even the weakest approaches to sustainable development (Kunstler, 1994). Compounding these antiquated laws and regulations are local power structures and economic power brokers (such as developers, lawyers, realtors) who are representative of corporate and even transnational corporate capital. They comprise core elements of a loosely structured growth machine coalition (Molotch, 1976; Logan and Molotch, 1987), and their access to and influence on the operating political process bias the "rules of the game" heavily in favor of the top dogs over the "semisovereign people" (Schattschneider, 1960). The supplanting of the implicit "rules of the game" of the local "growth machine" by the "rules of play" of the Sustainable City Game will be the fruits of hard-won victories of forces aided by vigorous grassroots leadership, educational initiatives and reforms, serendipitous opportunities, a mobilized public and the permeation of ecological values over time. Without exaggerating their impact, this change will also be the result of the growth of civic professionalism and ecological consciousness in expert communities and the efflorescence of new ideas and novel strategies for redeeming the promise of the city as the home of humankind.

As a modest contribution to a formidable challenge and a long-term goal, this chapter has offered a more radical ecological answer (radical, in the sense of going to the roots) to Castells' question of the shape that the new information city would take to provide "a new socio-spatial structure…made up of a network of local communes controlling and shaping a network of productive flows" (Castells, 1989: 353). Aided by the new rules of play of the Sustainable City Game, that new information city is the sustainable city.

Part II

SUSTAINABLE CITIES AROUND THE WORLD

Chapter 7

URBAN DREAMS OF GLOBAL SUSTAINABILITY

Introduction

As the era of late modernity fatefully slouches into the second decade of a new millennium, at least two contradictory responses to mounting evidence of increasing global environmental damage are apparent. On the one hand, the nations of the north, abetted by various international agencies like the United Nations and the World Bank, produce ringing statements of endorsement for the pathway of sustainable development and formulate protocols setting limits on the most worrisome problems affecting them. On the other hand, many nations of the south in alliance with environmental and peace and justice nongovernmental organizations (NGOs) advance locally based, nonhierarchical policy solutions that assign heavy responsibilities to the industrialized north to share in the burden of charting a new course beyond unsustainable practices and processes and toward global equity and environmental repair.

If the world is changing faster than academicians and policymakers can reconceptualize it, it seems that an acute problem of coming to grips with persistent problems of population control, sustainable development and international peace may lie in the difficulty of formulating a new vocabulary and conceptual framework. These would be adequate to grasp the complexity of the forces shaping key developments in the economics and politics of global change and their impact upon regions and localities trying to adjust or respond to these imposing elements. In debates over sustainable development, the concept of sustainable development remains contested, often mustering wide support and weakly-grounded consensus precisely because it is left unexplicated and not concretely defined (see Chapter 1). So too is the persuasive force of catchwords and principles of sustainable development purchased at the expense of clarity and specificity. Globalization, economic restructuring, international competitiveness, regional integration – these and other terms populating the policy language of the new international political economy carry much rhetorical force. This is not because their meaning is transparent and well understood, but because they, like myth, are poorly comprehended and serve as condensations for vaguely felt phenomena being widely sensed. No less so, the adage, "Think Globally, Act Locally," the rallying cry of many sustainability groups and organizations throughout the world, gains its power from its obscurity.

In part, this chapter – like Chapter 3 – illuminates the costs exacted by this conceptual unclarity and the uses to which it has been put by various schools of thought and powerful organizations. We begin in the first section with an exercise that explores the strategic implications of linguistic clarification of key terms – the "local" and the "global" – in the sustainability debate triggered by key participants in that policy contest (Wendell Berry, Alan AtKisson and Wolfgang Sachs). Then, we offer a critique of the deployment of those terms in the strategy of "green globalism" – a program identified with the policies of many nations of the north and of those of international organizations dominated by the industrial nations. This critique transforms in the second section into a critical examination of the perils of environmental moralism expressed in the proliferating number of international sustainability declarations enunciated at international conferences and multilateral meetings. Most particularly, we contrast the favored strategy of sustainable development growing out of the local implementation of the imperatives of Chapter 28 of Agenda 21 with the Aalborg Charter. In the third section, we critically examine four representative sustainable community programs in various parts of the world to determine the extent to which the tacit model of sustainable development championed by the Local Agenda 21 process has borne fruit.

Finally, the chapter concludes with an effort to recuperate the meaning of sustainable development beyond the false polarities between localism and globalism by showing how an alternative strategy of sustainable cities limned earlier in this work must combine the double pursuit of urban sustainability and global equity.

The Local(ist)-Globalist Debate

Ever since "Think Globally, Act Locally" became a rallying cry for a large segment of the international environmental movement, this adage has been the focus of sustained debate by varying quarters in the global controversy over sustainable development. Some, like Wendell Berry (1989), have taken the insistent position that only local action is productive, while global thinking is either incomprehensible as a component of restoring the built environment and the more encompassing ecosystem to health or worse yet, absolutely futile as a political strategy and social plan for action. Others, especially those who have striven to implement sustainability principles into local practices, have taken a decidedly different tack, pressing the thesis that as an integral feature of strategic politics, global theorizing is useful and even necessary. Alan AtKisson

(1990), an architect and guiding hand in the Sustainable Seattle program, has demurred from Berry's skepticism, arguing the utility of such global thinking. Working to adjudicate this friendly quarrel between environmental camps, Wolfgang Sachs (1992), a recognized member of the Wuppertal Institute, denied the fruitfulness of such antagonistic interpretations and has urged combatants to refuse to choose.

The import of this verbal sparring certainly goes beyond simple or simplistic renderings of these divergent positions. It opens into the pitched struggle among subaltern groups and marginalized elements around the world to undercut the hegemony of "green globalism" being articulated in multiple forums from the United Nations to the World Bank. From the United National Conference on Environment and Development (UNCED) to Agenda 21, a strategy of global sustainable development has been advanced by major political power brokers and economic and financial stakeholders bent on incorporating the term "sustainable development" into a global discourse that tames the concept, skews the agenda for action and change and valorizes a particular kind and construction of knowledge congenial to prevailing social relations and the distribution of power among key contestants in the debate.

This critique of "green globalism" put forth by a variety of scholars, critics and ecological activists (see, e.g., Shiva, 1993; and Lohman, 1993; and, generally, Sachs, 1993) seeks first of all to devalue the notion of "the global" from its privileged status on a largely uncontested intellectual terrain, marking it off as a keystone of a hegemonic discourse. A particular reading of "the global" has been granted authoritative status both by a historic bloc composed of leading groups within international political, banking and transnational organizations and by subordinate economic groups and social organizations whose ecological concerns, bases of knowledge and preferred outcomes from such policies run counter to the overriding consequences of the dominant reading. Precisely because of this, the politics of local/global conceptualizations and relations are central to the advancement of a counterhegemonic strategy that deprivileges "green globalism."

In this process of deprivileging, this alternate, counterhegemonic discourse removes the global from its vaunted and unchallenged position by identifying the prevailing construction of the global as a historically specific and politically and economically constructed institutionalization of a globalized local (Shiva, 1993). It also recognizes the particular way in which ecological issues rendered as "global environmental problems" are part of a hegemonic discourse involving distorted communication process veiling their deeper structural antecedents that promotes a highly truncated and biased agenda, which marginalizes key issues and concerns of local peoples and communities of the south (Rich, 1994). The limited focus and scope of the Rio Declaration and protocols testify to the largely successful effort of forces working through the medium of this United Nations forum to push to the periphery wider and more fundamental issues given voice but little media coverage in counterforums surrounding UNCED and World Bank events.

As these intellectuals and movement scholars (especially Lohmann, 1993) point out, even the globalist policies and programs for protecting biodiversity remain continuous with the impulse of technological modernity to treat Earth as a "standing reserve," now seen as precious and depletable, but whose urge to dominate is only modified by the necessity of rational eco-technocratic administration and placed under the eco-Taylorist regime of natural resource productivity. Thus, the UNCED protocol promoting biodiversity stands less as a manifestation of growing global ecological consciousness than as a testimony to globalist ambitions to enforce tighter and more rational methods of managerial-technocratic control over the exploitation of Earth's limited fund of natural resources. Such actions also point to the industrialized north's strategy of shifting blame and responsibility for ecological degradation from the north to the south.

Finally, this critique labels the discourse of "green globalism" as a totalizing discourse in that it rests squarely upon the Enlightenment foundations of Western epistemology. Here, expert knowledge, professional culture and scientific practice are pitted against local knowledge(s), so-called traditional culture(s), and ecologically and communally informed practice(s), and the latter are all found wanting. Since the discourse of "green globalism" is still at its base a modernizing discourse, its totalizing character is self-evident. As such, "green globalism" remains resistant to the fact that *sustainability is at base a critique of developmentalist ideology, not an extension or moderation of it.*

Against these tendencies toward globalist hegemony over this discourse, these critics proffer the local as the true ecological and political space of global ecology and sustainable development. In simpler framings of the primacy of the local over the global, this approach merely inverts the bipolar relation. But even more complex renderings leave largely unexplicated the character of the linkage between local and global adumbrated in the "Think Globally, Act Locally" aphorism (Henderson, 1988). In the section that follows, we will take up one vehicle for articulating this relationship: the formulation of international sustainability declarations. In this context, we will note the problems and perils of if this process, highlighting the risk of environmental moralism implicit in such practices.

Between Environmental Moralism and Policy Incrementalism: Local Agenda 21 vs The Aalborg Charter

Treaties and protocols have not been the only products of international conferences. Another type of document frequently arising from these meetings is an international statement of vision and principle that delineates vistas on the future, guiding principles and ultimate goals, plans of action and (sometimes) indicators of progress. Whether in the form of official reports, multilateral declarations, a checklist of principles, an agenda or a charter, these documents typically reflect the broad framework of those who promulgated the international meeting and the consensus reached in the dialogue and give-and-take sessions culminating in the end-product.

As more narrowly-framed environmental issues in the international arena gave way to the inchoate dialogue over sustainable development, various international bodies were convened or formed commissions to enunciate guidelines and goals for advancing sustainable development or the more generic term, sustainability, as part of a global or multilateral plan of action. An early example and important bellwether of this trend in international policy articulation was, of course, the United Nations-sponsored commission headed by former Norwegian prime minister Dr Gro Harlem Brundtland, whose efforts were brought to fruition in a report titled "Our Common Future" (UN, World Commission on Environment and Development, 1987). This influential report, as we have seen, has been credited with offering a definition of sustainable development that has become the touchstone for statements and declarations by other official bodies and organizations.

A careful assessment of many of these sustainable development documents will reveal a powerful tendency of such statements to become caught between the poles of environmental moralism and policy incrementalism. That is, on the one hand, these international policy enunciations on sustainability are often dressed up in the language of oughts, growing out of a scenario of the present and near future and caught up in an almost apocalyptic vision of imminent ecological degradation, catastrophe or doom. The latter, taking Rachel Carson's *Silent Spring* (1994 rep.) as its model, is epitomized in the Club of Rome's foreboding work, *The Limits of Growth* (Meadows, 1972). Yet curiously this eco-apocalyptic forecast is routinely followed, on the other hand, by policy recommendations couched in "oughts" and "shoulds," which are sometimes directed to nations of the south, who are encouraged not to follow the north's ecologically evil ways, but on other occasions are targeted to the nations of the so-called developed world who are called upon to take incremental steps that will lead in an additive way to the transformation of quantity into quality.

The problem with this orthodox strategy of development, as we have argued (see Chapter 1), is that these statements mistakenly focus policy action at the scale of the nation or entire globe. Targeting sustainable development strategically at the global or national scale all but makes impossible meaningful ecological/social analysis and concerted political action. Weighing and assessing dynamic and conflicting social processes and tendencies at the national or global scale contributing in some cases to sustainability and in others to unsustainability would prove to be a daunting if not impossible task and certainly one with wide margins of uncertainty, if not unreliability. On the one hand, the apocalyptic tone of such statements often mobilizes among mass publics a sense of fatalism or political paralysis in the face of such ominous forecasts. On the other hand, these analytic imponderables tend to foster a lack of clarity on the impact of national or global programs in affecting the sustainability quotient (the net balance or outcome of aggregate sustainable and unsustainable activities and trends) at the national or global scale. In addition, as the scientific debate surrounding the Kyoto conference on global warming demonstrates, global modeling techniques can lead to wide variance in predictions about the nature of the threat and the likely consequences of policy intervention upon issues like future climate change.

Amidst political wrangling over and media coverage of the fate of various treaties at the Rio Earth Summit, one of the outcomes accorded less attention and analysis was the conference endorsement of Agenda 21, which took place on 4 June 1992. Agenda 21 is a fairly extensive document, covering a range of topics fostering a new global environmental agenda in its 39 chapters (United Nations, 1992). In general, Agenda 21 was conceived by its originators, the Canada-based International Council for Local Environmental Initiatives (ICLEI) and national municipal organizations, as a mechanism for transferring the ideas and concerns voiced in global forums on sustainable development (like UNCED) into local programs for realizing community sustainability. In fact, its major significance for the sustainability movement worldwide has rested in the proliferation of activities stemming from a single chapter (Chapter 28: Local Authorities' Initiatives in Support of Agenda 21), specifically its call for local governing bodies in each nation to institute a consultative process culminating in the design and implementation of a "Local Agenda 21."

Reacting negatively to the globalist thrust of orthodox approaches to sustainable development, Local Agenda 21 swung to the opposite pole and inaugurated a localist strategy. The defining features of Local Agenda 21 planning identified by ICLEI and its local practitioners as critical to its success are:

• Multisectoral engagement in the planning process through a local stakeholders group, which serves as the coordination and policy body for preparing a long-term sustainable development plan;
• consultation with community groups, NGOs, business, churches, government agencies, professional groups and unions in order to create a shared vision and to identify proposals and priorities for action;

• participatory assessment of local social, economic and environmental conditions and needs, and;
• monitoring and report procedures, including local indicators, to track progress and to allow participants to hold each other accountable to the action plan (ICLEI, 1997b).

These elements comprise the guidelines for the more than 1,800 local governments in 64 nations – including Canada and the United States – carrying out Local Agenda 21 programs.

If the grounding principles and directing guidelines of Local Agenda 21 are founded on certain North American cultural assumptions and biases, the underpinnings of the Aalborg Charter (European Commission, 1994) reflect sometimes subtle, sometimes distinctive European cultural anchor points and emphases. This sustainable cities and towns charter grew out of the European Conference on Sustainable Cities and Towns, convened by the City of Aalborg, Denmark and the European Commission in Aalborg from 24 May to 27 May 1994. Over six hundred individuals and representatives from cities and towns, international organizations, NGOs and scientific bodies from Europe and elsewhere attended the conference, which culminated in the signing of the document by over ninety local governmental representatives and in the launching of the European Cities and Towns Campaign.

The identity and uniqueness of the Aalborg Charter, vis-à-vis Local Agenda 21, have been obscured in part by the directing role of the ICLEI in preparing the program of the conference and by a tendency to blur the different touchstones and underlying visions of achieving sustainability of the document. Both before and after the conference, ICLEI maintained the position that the conference and campaign were animated by the objective "to promote development toward sustainability at the local level through Local Agenda 21 processes" (ICLEI, 1997a). It also envisaged the achievement of sustainability as emerging from a lengthy passage by individual local authorities through a series of steps or actions leading to a condition of sustainability.

The substance of the Aalborg Charter was, however, heavily influenced by the key assumptions, guiding principles, core understanding and animating the Center for Sustainable Cities and underpinning its founding documents (Yanarella and Levine, 1992b and c). Drawing from this alternate strategy of sustainable cities, the charter enunciated a more complicated balance-seeking process of conceiving sustainability that eschewed the implicit pathway metaphor of the former.

In advancing the case for an alternative strategy of social sustainability or sustainable cities, the charter argues for a second and alternative image of sustainability as a balance-seeking process. This position advances the idea that urban sustainability requires a threshold involving a minimum set of conditions such that each succeeding step becomes easier, less costly and more politically tractable. Between globalist and localist strategies, the strategy of sustainable cities locates the city as the place and nexus for striving to put into balance the many levels of sustainable and unsustainable systems flowing through the city.

Case Studies in Sustainable Community Development: Status and Prospectus

UNCED's call to implement Local Agenda 21 programs, as noted earlier, was taken up by hundreds of local authorities around the world. The implementation process itself was developed and refined, not only by local governments and coordinating groups, but also by the International Council for Local Economic Initiatives (ICLEI homepage). In association with the United Nations, ICLEI established the Local Agenda 21 Model Community Program. Its purpose was to stimulate new sustainable community initiatives at the local level by highlighting the design and implementation processes of various programs and identifying "best practices" developed in these local enterprises. At its height, 19 programs circling the globe were chosen as model sustainable communities, and each became the focus of intense monitoring and periodic review. This section provides individual overviews of four representative model community projects from the United Nations-ICLEI program in order to assess their status and prospects. The overall purpose of this exercise is to evaluate the strengths and weaknesses of an implicit model of sustainable development informing the Local Agenda 21 movement.

Troyan, Bulgaria

Located at the edge of a biosphere reserve and national park, Troyan is a moderate-sized community with a population of 46,000 some 90 miles east of Sofia in the northern foothills of the Balkan Mountains. Its main industrial activities include the manufacture of pharmaceuticals, textiles and electric motors, highland agriculture and animal feedlots as well as logging and wood processing. As part of the heritage of the Stalinist economic command economy (heavy industry plus the pillage of natural resources), these economic enterprises contributed to polluted water, pervasive air pollution and other persistent environmental problems in the city and region.

In the transition from a political system characterized by central planning and top-down administration to one keynoted by representative governance and local democracy, Bulgaria worked to transfer environmental planning and management expertise to municipalities and localities like Troyan. In consultation with the United States Institute for Sustainable Communities, Troyan undertook an environmental action plan with the twin aims to promote municipal knowledge and expertise in environmental stewardship using a planning model incorporating environmental risk assessment and participatory decision-making methods. This program was named the Troyan Environmental Action Program (Troyan EAP).

In its efforts to shake off the barnacles of state socialism, the municipality became involved in activities and programs to:

- Survey a sizable sample (5,000) of local citizens to determine the ranking of environmental problems (with water pollution topping the list);
- generate a comparative risk analysis utilizing the best available scientific data on the health, ecological and socioeconomic risks associated with these publicly prioritized environmental problems;
- develop action plans to remedy these problems in the most cost-effective ways; and
- establish the Troyan Environmental Education and Information Center to disseminate wider public understanding about evolving environmental renewal and sustainability programs (ICLEI, Case Studies on the Local Agenda 21 Process: Troyan, n.d.).

Initiated in 1989, Troyan's 21-month EAP resulted in a number of achievements. In the first place, the Troyan EAP became something of a model of environmental planning and management in this Central European nation employing democratic decision-making tools for other Bulgarian municipalities. Secondly, new municipal laws and regulations were promulgated during the program's time period that improved access to information and clarified the rights and responsibilities of the local administration, citizens and companies with regard to environmental protection. Thirdly, the relationship between the municipality and industry with respect to environmental problems improved considerably in comparison to Bulgaria's Stalinist era, as firms and industries were required to submit detailed reports and obtain appropriate permits regarding environmental impacts. The municipality itself undertook programs to redesign solid-waste and wastewater treatment policies and to institute water conservation programs.

The prevailing record also indicates how far it departed from a common Soviet and East European experience typified by the Communist Party's Promethean faith in high technology (recall Lenin's definition of communism as the "Soviets plus electrification" (1966: 516) as well as his fascination with

Fordist mass production) and its reckless disregard for the social and human health consequences of despoiling nature and ravaging its treasures. On the other hand, the character and status of the Troyan EAP vis-à-vis any model of strong sustainability showed how distant it remained from a genuinely integrated, focused and holistic sustainable cities program. Its major thrust was grounded in an admittedly participatory but largely liberal environmentalist agenda to repair some of the worst of the exploitative industrial-technological practices of a prior authoritarian state socialist regime and its national enterprises.

Municipal Durban, Kwazulu/Natal Province, Republic of South Africa

Durban is located on the east coast of the Republic of South Africa. With an average growth rate of 2.8 percent per year, its 2.2 million population was expected to exceed 5 million by the year 2000. With an apartheid legacy of deep social cleavages and the worst political conflict and internecine violence experienced in the country, Durban is one of the most challenging of the urban areas in the nation's struggle for social and economic reconstruction.

The first city in South Africa to formally respond to the Local Agenda 21 mandate, its sustainable community initiative – known as the Sustainable Reconstruction and Development program – was designed to ensure that at the local level, environmental considerations would become an integral part of the high-priority Reconstruction and Development Program (RDP) of the new South African government. The RDP provided a policy framework for national development to address issues of poverty and inequality through economic growth. Concerned that environmental needs not be nullified or marginalized in this quest, Durban's Local Agenda 21 program sought to promote sustainable development and integrated environmental management into the new system (ICLEI, Case Studies on the Local Agenda 21 Process: Durban, n.d.).

In its initial phases, it emphasized establishing new municipal administrative structures and processes to support the Local Agenda 21 planning effort. This involved restructuring internal departments to reflect the long-term sustainability agenda. Beyond that, it facilitated interaction among the city, local research groups and NGOs, which work through a widely representative steering committee convened to oversee and direct the State of the Environment reporting process.

Since its inauguration in August 1994, Durban's Local Agenda 21 process had to confront the continuing heritage of apartheid as a millstone retarding the policy ambitions and efforts to realize its dream of sustainable development.

On the positive side, it was largely successful in integrating its Local Agenda 21 into the initiatives of the Department of Urban Development, Department of Urban Strategy Development and Department of Parks as a keystone of its national reconstruction and development strategy. These efforts were reinforced by important strides in involving key corporate and community elements as stakeholders in the Local Agenda 21 process. In so doing, an important political dividend of this stakeholder process involved the forging of a consensus-building process contributing to a culture of participation and community support. Still, the problems and dilemmas thrown up by the legacy of apartheid continued to be felt as social inequality and poverty weighed down efforts to achieve a more equitable program of sustainable development for all citizens.

City of Santos, Brazil

Situated on the Atlantic coast of Brazil, Santos is a small municipality with an annual population growth rate of around 2.8 percent. While its industrial base is almost nil, it is located close to the metropolitan region of Sao Paulo, whose population exceeds 19 million. Thus, while two-thirds of its local economy centers on the tourist industry, it is the gateway to a vast and highly industrialized region to its west. With a population approaching 429,000 when its program began, Santos was confronted with a host of contradictory and knotty political obstacles besetting its ambitions to realize its dream of sustainability.

In the context of a history of military dictatorship and the curbing of civil rights, economic crises and environmental degradation, Santos adopted an approach involving popular participation in municipal policymaking and programs intended to build the foundations of a sustainable future. The institutional vehicles for advancing its municipal sustainability agenda were the Citizens Councils for Public Participation, instituted in the early 1990s. Their specific role in this enterprise was to involve service users in the definition, management and implementation of municipal services (Knight, 1996).

The Citizens Councils for Public Participation were surprisingly successful in carrying out an agenda intertwining social and ecological objectives. Their guiding strategy was to promote widespread community education and political participation by, among other things, demystifying scientific knowledge and practical technology. In deprivileging status of technical know-how, these citizens' councils mobilized and steered the collective intelligence and good common sense of popular elements and assemblies into more open democratic policy processes. One important dividend of this combined social justice and political education campaign was the palpable rise in the standard of living of hitherto marginal groups and strata of the community.

Another was the heightening of popular awareness and understanding of the embeddedness of the community in the natural ecology and the interdependency of the community's quality of life with the health and well-being of the environment.

Illustrative of this dual educative process was the restoration of the Bugre River ecosystem and the promotion of its Dique Project, which has involved the construction of new housing for nearly 20,000 residents under the labor-intensive banner, "more workers and fewer machines" (ICLEI, Case Studies on the Local Agenda 21 Process: Santos, n.d.). While class inequalities persisted and paternalistic relationships between government and citizens remained, rejuvenated bottom-up activities like those enumerated above began to erode old practices and relationships associated with Santos' colonial past (ICLEI, Case Studies on the Local Agenda 21 Process: Santos, n.d.).

Operating under the fundamental principle of popular democratic administration, the municipality engaged in a variety of social and environmental programs to transform Santos into a sustainable community. An especially high-priority program was its effort to elevate the quality of present housing stock and establish basic infrastructure for residents living in substandard housing and in environmentally sensitive regions. Another important item involved the development of a program of biodiversity conservation through a number of policy instrumentalities, including environmental education and eco-tourism. Because of the extent of wealth and income inequity in the area, creating and augmenting social services for poor residents in the old colonial center was another important and well recognized activity of Santos' Local Agenda 21 program. The revival of the old colonial center was furthered through the building of new coalitions and partnerships among key stakeholders in business, the citizenry, academia and governmental agencies. As a result of the role and importance of Santos' beaches in its sustainable development strategy, improvement of area water quality through control of solid and liquid waste pollution of its oceanfront was emphasized. Finally, given its focus on issues of social justice, Santos' sustainable community program targeted the improvement of the quality of life of peripheral social elements in the city (street people, abandoned children and psychiatric patients) for innovative social services.

Through forums, commissions, nuclei of defense, public meetings and councils, this Local Agenda 21 enterprise became dedicated to bringing into public life those sectors of society that had historically been excluded from decision making (women, children, the unemployed, working poor and unionized workers, etc.) organized around weaving a sustainable web incorporating social and democratic renewal and ecological sustainability (ICLEI, Cases Studies on the Local Agenda 21 Process: Santos, n.d.).

While Santos' checkered record combined a genuine commitment to grassroots participation in local environmental planning with a strong community agenda advancing social justice, it also embraced a weak or minimalist definition of sustainability that accommodated the productivist ethic of industrial development characteristic of earlier theories of economic development, primarily favored by First World nations and international institutions like the World Bank. Its legacy also points to the dilemmas of a local community influenced by left-progressive ideological ideas and platforms in trying to synthesize economic development, social justice and ecological stewardship.

Hamilton-Wentworth, Ontario, Canada

The Canadian commitment to sustainable development has taken greatest expression in the many programs for local sustainability populating its landscape. Among these myriad programs is the Vision 2020 Sustainable Community Project of the Regional Municipality of Hamilton-Wentworth. With a population of around 452,000 (Statistics Canada, 1991), the Regional Municipality of Hamilton-Wentworth is located in the southwestern part of Ontario along the western shores of Lake Ontario. When economic restructuring caused many of its largest steel and related manufacturing companies to downsize, relocate or even close, public officials, urban planners and corporate executives worked to rebound the regional economy on high-tech industries, environmental produce, service firms, and increased value-added businesses.

Until recently, Hamilton-Wentworth's Vision 2020 was regarded as the crown jewel of Canada's many sustainable community experiments, having been chosen in 1993 as one of nineteen or so communities throughout the world in ICLEI's Local Agenda 21 Model Communities Program. Its origins lay in the concatenation of management and planning imperatives and regional environmental concern taking place in late 1989 and early 1990. During this period, a regional management and planning review coalesced around the concept of sustainable development as the appropriate anchor of a new regional planning and budgeting system. The other event hastening the development of the sustainable community initiative was grassroots organizing around the issue of the mounting pollution of Hamilton Harbor (Yanarella, 1998).

From these stimuli, a citizens' task force was instituted, which from 1990 to 1993 engaged over a thousand citizens in lengthy meetings, interviews, focus groups, round tables and town sessions. The outcome of this mutual public education campaign was a community vision statement titled "VISION 2020: The Sustainable Region," which delineated the outlines of a sustainable Hamilton-Wentworth in the year 2020 based upon the principles and requirements of sustainable development. Following close on its heels, an indicators project called "Signposts on the Trail to Vision 2020" launched in the summer of 1994 identified 24 key measures of sustainable development for the region, which were used subsequently to monitor and assess progress toward sustainability (ICLEI, Case Studies on the Local Agenda 21 Process: Hamilton-Wentworth, n.d.; and Regional Municipality of Hamilton-Wentworth, 1996a).

Incorporating nearly 100 of the 400 proposals listed in Vision 2020, the Regional Council passed a new Official Plan for Land Use titled "Toward a Sustainable Region" in June 1994. This was followed in November 1994 by adoption of the "Renaissance Report," establishing a strategic plan for long-range sustainable development. A Regional Council staff working group subsequently designed vehicles for integrating sustainable development principles and the vision statement into the formulation of capital budget and departmental work programs (Regional Municipality of Hamilton-Wentworth, 1996b and c).

Beyond these bureaucratic and administrative innovations, Vision 2020 gave rise to a host of specific projects, ranging from development of a bicycle commuter network to alleviate automobile transportation loads and a Hamilton-Wentworth green-up program to assist families and business in implementing conservation and other sustainability measures to construction of combined sewer overload reservoirs to improve harbor water quality. It also sponsored an annual Sustainable Community Day beginning in 1994, where public education continued to be the focus.

Hamilton-Wentworth's Vision 2020 Sustainable Communities program stood out for its extraordinary success in balancing governmental and administrative prerogatives and actions with democratic consensus decision making and direction. Regional planners, business people and venture capitalists also demonstrated an uncommon ability to discern creative ways of greening the local economy and confronting the process of economic restructuring affecting the region (Kendrick and Moore, 1995).

While these developments have been promising, Vision 2020 has had to weather dramatic shifts in provincial support over its relatively brief tenure. Beginning with strong support from a sympathetic New Democratic Party provincial government, it suffered from diminished political and financial assistance by the Mike Harris Progressive Conservative victory and its "Commonsense Revolution" in Ontario. Sizable reductions in budgetary support for government services and programs have undercut the planning ambitions and goals underpinning Vision 2020 and the Renaissance Report, as well as weakened conservation authorities and other environmentally-related measures (Prokaska, 1995).

A more symbolic but in other respects more formidable attack on Vision 2020 came from the opponents of regional plans to build a north-south expressway through the Red Hill Valley. Foes of the Red Hill Creek Expressway argued against the highway construction project on the basis of its incompatibility with some of Vision 2020's principles and design goals (Lukasik, 1996). This political controversy points to an even more fundamental shortcoming in Vision 2020's planning and implementation to date: downtown revitalization and its relationship to transportation policy. Although the program focused attention on the revitalization of the downtown through activities such as a Downtown Ideas Charette (Mokrycke, 1996) and the municipal government-commissioned transportation policy study, little headway has been made in finding meaningful ways to curtail the automobile as the primary means of transportation in and out of the downtown business area, or to curb its role in shaping the form and pattern of development of the downtown or suburbs. Preponderant attention has been given to alleviating anticipated traffic bottlenecks through shopworn traffic nostrums, rather than by giving priority to pedestrian, bicycle, and public transit and attendant recommendations flowing from a genuine vision of a more compact, densely populated sustainable city of the future.

Looming on the horizon like a slow tidal wave barely perceptible in the distance have been the negative shaping forces of growth pressures from the Greater Toronto Area (Poling, 1995). Unchecked expansion of the Greater Toronto Area has continued to threaten to engulf the regional municipality and submerge its distinctive regional identity into a more homogeneous urban agglomeration. Thus far eschewing a strong sustainability platform for its thin sustainability approach, the Hamilton-Wentworth program has historically seemed to satisfy itself with focus and action on the parts (e.g., its valorization of sustainability indicators), while leaving the vision of the whole to languish from inattention and neglect.

Critical Evaluation

These four case studies of model sustainable community programs scattered around the world point to the promise and limitations of the Local Agenda 21 process. On the positive side, they exhibit the enormous variety and distinctive talent and inventiveness of local-based programs responding to a global agenda. While Chapter 28 of Agenda 21 provided the inspiration for these local programs, the distinctive and idiosyncratic elements of these cities and towns as socially and historically rooted places provided the particular resources for generating and carrying forward these programs.

The local focus of these Local Agenda 21 programs was another positive feature of this local-global movement. As the critique of orthodox strategies of sustainable development discloses, globalist pathways to sustainability threaten to be swallowed up or engulfed by the many countertrends toward unsustainability operating internationally. Thus, a balance sheet weighing the aggregate sustainable practices and their impact against formidable unsustainable trends poses itself as a virtually insuperable exercise. As the "Sustainable Cites Manifesto" (see Chapter 3) and the Aalborg Charter argue, it is the scale of the city itself that is the nodal point for social synthesis and the platform for productive change seeking to realize sustainable development.

On the other hand, despite ICLEI's efforts to monitor these projects since their inception and to evaluate critically their progress, the vagueness of the original call for the proliferation of Local Agenda 21 programs and the looseness of its criteria for assessing progress toward sustainability have vitiated this enterprise. Lewis Carroll is often cited as the author of the expression: "If you don't know where you are going, any road will get you there." The character and direction(s) of these (and other) model sustainable community programs clearly fell vulnerable to the spirit of this adage. In the first place, these many local roads to sustainability left suspended the meaning and nature of the putative objective of their programs – sustainability. In a sense, each was a collection of focused projects whose goal was really to reduce unsustainability rather than to achieve sustainability. The conventional tool of generating sustainability indicators is really an instrument for measuring movement away from unsustainable activities or conditions, not a tool for assessing progress toward sustainability.

In advancing the case for an alternative strategy of social sustainability or sustainable cities, we have insistently argued for a second and alternative image of sustainability as a balance-seeking process. Between globalist and localist strategies, the strategy of sustainable cities locates the city as the place and nexus for striving to put into balance the many levels of sustainable and unsustainable systems flowing through the city.

Though grounded in the key insight that sustainability is best achieved at the local levels, such programs acted as if such policy incrementalism resting on reducing unsustainability could be magically transformed into a qualitative endpoint where sustainability crystallizes and become entrenched. These programs also carried along the mistaken notion that local enterprises could be hermetically sealed from wider regional, national or global forces. The ongoing local-regional-national and global transactions continuously interacting with the local fly in the face of the autarkic state presumed by the Local Agenda 21 model. By contrast, our alternative model sees sustainable cities as nodal points of these interactions and seeks to incorporate them into the operation of this balance-seeking process model of sustainability.

While the first step or phase in achieving sustainability (i.e., moving toward sustainability) may be necessary to muster the political conditions for realizing the second step or phase of achieving sustainability (i.e., instituting a balance-seeking process), the first must never be mistaken for the second. For sustainable development or sustainability to be meaningful and have substance, it must transcend the pathway metaphor and its quantitative/incrementalist approach for the balance-seeking metaphor and its qualitative/holistic framework. However meritorious were these halting steps toward slowing or reducing unsustainable practices in their communities, these Local Agenda 21 model sustainable community programs failed to enlarge the scope of their activities to encompass the demands and requirements of the balance-seeking process model. Lacking that second step, they could not truly move beyond Local Agenda 21 to the Aalborg Charter and its underlying balance-seeking process framework.

Conclusion

The animating thesis of this chapter is the underlying argument that neither sustainable cities nor a new international economic ordering grounded in sustainable practices can be achieved without the other. At base, sustainable cities require global equity, while global sustainability can only be truly realized if north and south find a new economic framework to undertake a transition to genuine sustainable development.

Under today's iniquitous power relations – both national and global – the pursuit of a meaningful and vigilant campaign to catalyze the building of sustainable cities worldwide and the transformation of north–south relations beyond the confines of the ideology of the market and the power of transnational corporations and their supportive international organizations must seem daunting, if not simply utopian or pie-in-the-sky. Yet beneath and around the constellations of power, finance and privilege heavily shaping the global political economy are myriad NGOs, local-global movements and coalitions pressing for an alternative agenda for the new millennium. Liberal reformist and eco-technocratic strategies can perhaps soften and cushion some of the social and ecological consequences of a market-driven and transnational corporation-steered approach to sustainable economic development of the orthodox sort. But if indeed the arguments of eco-economists like Herman Daly are correct, the alternative model of sustainable development acknowledging the rootedness of even the international economy within a more encompassing ecosystem may yet find its greatest ally in the ultimate intractability of nature to continued instrumentalist-extractive assaults on its resources and its limits.

The twin themes of the limits of nature and the revolt of nature have been favorite ones of twentieth-century critical theorists (e.g., Horkheimer and Adorno, Marcuse and Bloch) who have sought to overcome the constraints and blind spots of increasingly shopworn nineteenth-century ideologies and philosophies. The emerging and inventive praxis of twenty-first century social movements, inspired by the novelty of the times and the need to transcend sterile ideological disputes, points to the antiquated character of many past forms of political practice based on mechanistic and deterministic assumptions of political optimism. Grounded in a reasonable and open sense of hope, their political tactics and strategies remind us of what is truly at the core of the human spirit and what redeems humankind from that "arrogance of humanism" that David Ehrenfeld (1981) detected in every secularly grounded, progressive ideology emanating from Enlightenment foundations.

By returning to our roots and recognizing our fragility and dependency upon those relations with nature that both limit us and sustain our inhabitation of this world, a strategy advancing both sustainable cities and global equity may yet prove to be the only real, nonutopian alternative to a divided, conflict-ridden and ultimately ruined planet toward which the globe's Earth brokers are steering us. Yet this eminently practical and nonutopian goal, as Reid and Taylor (2010) have cogently and compellingly elaborated, can only be achieved by rehabilitating the commons and weaving together into a more potent coalition the many groups and voices striving to advance that task in so many disparate ways across the globe.

Chapter 8

THE PROMISE AND PITFALLS OF CHATTANOOGA'S ENTREPRENEURIAL "SUSTAINABILITY" STRATEGY

The sustainability city program in Chattanooga, Tennessee, which began in the early eighties, is one that attracted enormous public attention and comment by public officials, government bodies, academic analysts, policy researchers, journalists and other media around the nation. In general, it received extravagant praise for its efforts and accomplishments, being variously characterized as "crown jewel" or "belle" of America's national sustainability efforts and portrayed as a modern-day "Cinderella story" yielding a model strategy dubbed the "Chattanooga Process" or "Chattanooga Way." Indeed, scholarly case analyses and policy studies of the "iron city to greenest sustainable city" narrative have brought to this southeastern railway hub (historically symbolized by its iconic train, the Chattanooga Choo-choo) not only a raft of regional and national awards and recognition. It has put this city on the sustainability map as one of two dozen or so cities that "take sustainable cities seriously" (Portney, 2003). That this southeastern mid-sized Tennessee city should have achieved such a hallowed status in the annals of the short history of sustainability in the United States was neither self-evident to most Chattanoogans nor a matter of destiny, given its origins and circumstances in the tumultuous decades of the sixties.

In this chapter, we offer a case study of the Chattanooga sustainability program that begins with the officially-sponsored story of its rise and success constructed from mainstream promoters and supporters of this campaign who often had strong interests in telling this story in a manner that gave Chattanooga's efforts maximum positive exposure and played to their personal benefit. This narrative concentrates on the role of prominent political leaders, the local chamber of commerce, other key community stakeholders and a major regional foundation in setting in motion a Sustainable Chattanooga program dedicated to overcoming blight and rebuilding this southern city along lines that they would initially designate as environmental in import and then reframe as sustainable. The result, as we show, was an example of a corporate-led program that constructed a form of "Showcase Sustainability," which revitalized the downtown and significantly reduced local pollution but failed to take into consideration significant social equity issues that comprise the third leg of the so-called sustainability tripod (economic health, environmental protection and social equity).

Using the strong sustainability framework outlined in Part I of the book, this chapter then delves into the Chattanooga Mirage and delineates the underlying reasons for the yawning gap between image and reality and the sources of the failure of this success story to institutionalize strong sustainability at the city-region scale. From there, we point to the reactivation of so-called sustainability efforts in the middle of the first decade of the new century, showing patterns of continuity and innovation in this renewed campaign. A series of policy recommendations convergent with our earlier strong urban sustainability strategy and set of practices will be issued to demonstrate what would need to be performed to bring this latest local-regional effort into conformity with strong sustainability in order better to realize the sustainability dream.

The Sustainable Chattanooga Story – The Official Discourse of Its Decline and Revitalization

Historical and economic background

Chattanooga is geographically nestled between the Appalachian Mountains and the Cumberland Plateau in southeastern Tennessee along the Tennessee River approximately 100 miles north-northwest of Atlanta, Georgia. Prior to settlement, it was inhabited by Native American tribes, most notably the Cherokees. After displacing the Cherokees, white settlers took advantage of its river connection and helped turn Chattanooga into a major steamship center and a significant military post. One of the sorrier chapters in its early history was the forced migration in 1838 of the native settlers called the "Trail of Tears," which pushed the indigenous Cherokees and other tribes to emigrate to Oklahoma. By the 1850s, Chattanooga had become a major railroad hub, which contributed to the development of its tourist and commercial industries. Its present African-American population can be attributed less to the south's cotton industry and plantation economy which was stymied by Chattanooga's mountainous terrain than to the migration of blacks there as a result of a more favorable racial climate and better living conditions stemming from white paternalism, the need for domestic servants and the existence of manufacturing jobs. With the arrival of the Industrial Revolution to the American South by the late 1800s, tourism became supplanted by manufacturing as Chattanooga's leading industry as its local economy became known for its focus on textiles, iron and steel fabrication. (For this history, see generally: Eichenthal and Windeknecht, 2008.)

With its continued growth in the manufacturing sector through the early part of the twentieth century, Chattanooga began to experience the environmental consequences of its heavy industry. By 1969, the city had the dubious honor of being dubbed the worst polluted city in the United States by the United States Department of Health, Education and Welfare. Local remediation efforts totaling $40 million were necessary to bring airborne pollution levels down to meet local and federal requirements in 1972. When amendments to the national Clean Air Act were passed in 1977 imposing more stringent regulations for particulate pollution, Chattanooga found itself again in noncompliance with the new federal regulations (see, among others, Parr, 1998).

Compounding its environmental pollution problems, which extended to significant pollution of its waterways (including the Chattanooga and Chickamauga creeks) from toxic chemicals dumped into these water bodies, were the consequences of deindustrialization of the city and region, which took the form of many industrial closures and consequent job losses (some 11,000 between 1978 and 1985). This trend led to the decay and hollowing out of the downtown and flight of downtowners to the suburbs; it also prompted the shift toward lower-paying jobs in the service and retail trade industries. (In addition to Eichenthal and Windeknecht, and Parr, see: Glick, 1996.)

Chattanooga's Cinderella story[1]

Beginning in the decade of the eighties, a succession of events and decisions (heavily shaped by Chattanooga's business and philanthropic communities) sought to come to grips with Chattanooga's precipitate economic decline and pollution problems. Two significant events took place in 1980. First, the Lyndhurst Foundation, founded and funded by the Lupton family that made its fortune through the Coca-Cola soda industry, hired a team from the Institute for Environmental Action (IEA) to investigate the underlying sources of the city's downward economic spiral and environmental problems. Its report blamed many of its prevailing woes on communication problems between citizens and their elected leaders and representatives. The IEA report also pointed to the weaknesses in its civil

1 This official discourse has been reconstructed from the history of Chattanooga's fall and rise, which resulted from the many economic boosters provided by Chattanooga's urban-economic-environmental initiative. This includes: Eichenthal and Windeknecht, 2008; Glick, 1996; Marston, 2004; Parr, 1998; Pierce and Johnson, 1998; and Youtie, 1999.

society stemming from a poverty of community and neighborhood groups and few opportunities for social interaction or nightlife in the declining downtown area. In short, Chattanooga needed to elevate the quality and livability of its urban core.

The second impetus to change came from the formation of the Civic Forum Task Force sponsored by downtown businesses and the economic development council of the Chattanooga Chamber of Commerce. Formed to uncover the reasons for Chattanooga's industrial stagnation, the task force reached the conclusion that in comparison with other Tennessee cities, Chattanooga had become an "invisible city" (American Forum for Global Education, 1995: 1) lacking a distinctive image or – to use contemporary marketing language – a brand to either distinguish itself from other mid-sized cites or to attract new industry in an era of deindustrialization affecting other United States cities. Unless this former "Scenic City" (Associated Press, 1995: 1) worked to refashion itself into a livable city characterized by balance, variety and an improved quality of life, the task force concluded that Chattanooga risked plummeting even further into the economic margins and becoming a mediocre, nondescript entity.

These sobering diagnoses mobilized major stakeholders in Chattanooga's future to follow the lead of a number of organizations brought into being by the financial largesse and organizational leadership of the Lyndhurst Foundation and the local Chamber of Commerce. In quick succession, the Moccasin Bend Task Force was formed to develop a plan for Chattanooga's declining riverfront. Then, in 1984, Chattanooga Venture was organized to foster wider public involvement in community planning of Chattanooga's revitalization. As ideas and proposed projects for downtown redevelopment and economic renewal were elicited in visioning exercises, community forums and design charettes through facilitation by Chattanooga Venture, a project called Vision 2000 took shape as an estimated 1,700 citizens participated in these public meetings and formulated 40 goals to steer projects and programs to realize that vision. Then, between 1982 and 1984, the Urban Design Studio became a prominent force in implementing facets of the inchoate plans for downtown renewal. By 1986, the landscape of Chattanooga's city planning and economic development scene had been dramatically altered as the RiverCity Company emerged to implement development plans; the Chattanooga-Hamilton County Regional Planning Committee was instituted to explore the regional implications of the city's initiatives; and Chattanooga Neighborhood Enterprise was brought into being to advance urban renewal through the repair of substandard housing and the promotion of affordable housing.

Table 8.1. Major projects and results

Project	Cost	Results
Tennessee Aquarium	$45 million (aquarium) $14 million (IMAX theater) $19.7 million (infrastructure)	– 1.1 million visitors a year to the aquarium – $365 million private riverfront development – Rents on downtown properties have risen from $13 a square foot to $50 a square foot
Riverwalk	$25.6 million	– 1998 Outstanding Planning Award, American Planning Association (Riverwalk)
Walnut Street Bridge	$4.5 million	– 1995 Honor Award for Urban Design, American Institute of Architecture (Walnut Street Bridge)
Electric shuttle and garages	$30.7 million	– 1.5 million passenger trips in 1993 and 1994 – Businesses located along the shuttle route have shown significant increases in sales
Grand Hotel and other CNE projects	$150 million	– 4,481 units of housing (90 percent low-income)
Orange Grove Recycling Center	---	– Employs 110 disabled individuals – Reduced landfill loads, resells 500 tons of separated materials per month, increased community awareness
Chattanooga/ Hamilton County Business Development Center Incubator	$2.3 million	– 220 companies started business, 120 companies graduated, 1,400 total jobs created, $200 million sales generated – Won 1997 Business Incubator of the Year award (manufacturing and services category)
Futurescape	$100,000	– Developed new comprehensive long-range plan – Changed zoning ordinances to require minimal landscaping and allow small lot, open-space designs
Eastgate town center	$30 million	– Four service industry firms 4,000 workers – Occupancy exceeding 90 percent
Southside redevelopment	$28.5 million	– Under development
SMART Park	$150 million	– At the prototype stage

Source: Youtie, 1999: 42

As these activities overlapped and converged on the goal of a prosperous and revitalized downtown, the target of transforming Chattanooga into the "best mid-sized city in America" began to morph into a reformulation of these initiatives into a new brand: the image of Chattanooga as "the Environmental City" (Flessner, 2008: 1). As the city moved into compliance with federal ozone standards and sought to meet amended Clean Air standards, media focus on Chattanooga began touting its ambitious Riverpark, Tennessee Aquarium and Walnut Street Bridge projects along the riverfront (many of which were heavily funded by the Lyndhurst Foundation) in terms of a veritable environmental turnaround by the city. By 1991, the environmental reinvention of the city's brand was well at hand. A well-honed marketing strategy was spawned by RiverValley Partners, the city's lead economic development agency, both to recraft the city's environmental image as a clean and environmentally responsible city and to generate industrial recruitment funds from public and foundation sources seeking to convert Chattanooga into a magnet for environmentally related business enterprises.

A number of actions lent substance to this rebranding of Chattanooga during this period. Particularly noteworthy was the decision by the Chattanooga Area Regional Transit Authority (CARTA) in 1992 to seek bids from high-technology transportation companies for battery-operated electric buses. In part, CARTA was prompted to advance this initiative to help mitigate the airborne pollution from diesel-fueled buses constituting its fleet; in part, it sought to alleviate problems flowing from inadequate parking space due to the rising downtown tourist industry. When no bidders were forthcoming, CARTA looked to the example of Santa Barbara, California, whose transit authority forged a partnership with a local manufacturing company to build electric buses. Emulating the Santa Barbara partnership model, CARTA sponsored and subsidized the creation of a local firm, Advanced Vehicle Systems (AVS), which began producing electric vehicles initially for Chattanooga and then for other cities in a growing national market.

With the publication of the Brundtland Commission report, "Our Common Future," in 1987, the word "sustainability" gained increasing popularity and purchase in local, national, and international circles. In 1993, the Clinton administration established the President's Council on Sustainable Development (PCSD). Charged with the mission of advising President Clinton on bold, new approaches to achieve economic, environmental and equity goals (the tripartite objectives constituting sustainable development), PCSD quickly turned its attention to emerging models of sustainability among American towns and cities. In 1994, a PCSD staff member visited Chattanooga, riding on the CARTA downtown electric-power buses and touring the Tennessee Aquarium; the outcome of his visit was his conclusion that Chattanooga was an example of an American city on the cutting edge of urban sustainability. PCSD's executive director was sufficiently impressed by the Chattanooga developments that she was moved to say, "there's a small revolution going out there" (Glick, 1996: 2).

Between 1994 and 1996, Chattanooga entered into a highly productive synergy with President Clinton's sustainability development council. Having gotten on PCSD's radar, Chattanooga's movers and shakers began to subtly recast its downtown revitalization and environmental image from the "Environmental City" (Flessner, 2008:1) to the "Sustainable City" (Madison, 1997: 1). James Frierson, a lawyer, businessperson and director of strategic initiatives for RiverValley Partners, pointed to the role of outside assessment of the Chattanooga turnaround, saying, "People came in and said, 'You may not know this, but what you're doing is sustainable development'" (Frierson, 2003). As the imaging bandwagon mounted, Chattanooga found itself the focus of countless news articles and policy case studies touting its sustainability strides. In turn, Chattanooga became host to PCSD through Vice President Al Gore, former Tennessee senator, who brought its annual meeting of the PCSD to spotlight the Chattanooga example. Thus, the PCSD came to act as both catalyst and legitimizer of Chattanooga's changing brand. Around this same time, the City of Chattanooga put up a Sustainable Chattanooga website outlining the community's commitment to sustainable development and city council member David Crockett established the Chattanooga Institute with initial funding from the Lyndhurst Foundation to proselytize the Chattanooga visioning model to other cities across the nation. It was not long after that this Chattanooga-PCSD symbiosis led to the conferral of awards for its sustainability enterprise – including the PCSD's first ever Sustainability Award, the Urban Land Institute's Excellence Award, the United Nation's Man and the Biosphere Outstanding Organization Award, the New America National Award for Sustainability and a host of other accolades. By the end of 1996, Chattanooga business, nonprofit and civic leaders enshrined this shift in branding by announcing its Chattanooga Strategy for Sustainable Development, which was grounded in promoting environmental well-being, social equity and economic development as its hallmarks (Chattanooga Chamber of Commerce, 1996).

As the city's new projects for downtown redevelopment continued apace and organizations, alliances, committees and offices proliferated, Chattanoogans expanded the scope of their revitalization efforts and sustainability activities to encompass a county and regional vision into the new century that sought to address issues like urban sprawl, regional technological innovation, and brownfields cleanup and redevelopment.

Table 8.2. Major organizations involved in Chattanooga Renaissance

Organization Name	Function
Lyndhurst Foundation	A private foundation that provided seed money to Chattanooga Venture, RiverValley Partners and other projects
Riverfront/Downtown Planning and Design Center	A nonprofit, public-private partnership that developed designs for the implementation of the Tennessee Riverpark Master Plan with funding from RiverValley Partners, the City of Chattanooga, and the Lyndhurst Foundation, and students from the University of Tennessee's Architecture College
Chattanooga City Government	City government provides funding to RiverValley Partners and other organizations involved in sustainable development projects, and funds infrastructure for downtown improvement projects
Chattanooga County Government	County government shares the city government costs, and the costs of ongoing management, maintenance and security along the Tennessee Riverpark system
Chattanooga Area Regional Transportation Authority (CARTA)	The transit authority responsible for the electric-bus initiative, an implementation of community goals to improve transportation and reduce pollution
Chattanooga Neighbourhood Enterprise, Inc.	A private nonprofit organization that addressed community goals for improved low- and moderate-income housing by renovating old downtown buildings and providing financing to residents
Chattanooga/Hamilton County Regional Planning Agency (RPA)	The county government agency that develops land-use plans and zoning ordinances in support of sustainable initiatives and the community goal of high-quality community development
The Chattanooga Institute	A private consulting firm that promotes sustainable development projects

Source: Youtie, 1999: 38

Oddly, by 2003, not a single formal institution involved in city policy, business development or citizen action bore or carried the name "sustainability" or "sustainable development" – explained by James Frierson, now founder and director of the Kruesi Center for Innovation, as stemming from the fact that "sustainability had now so completely seeped into the DNA of the community that the term is no longer needed and is everywhere practiced" (Frierson, 2003).

The Chattanooga story as mirage: The counternarrative

The dominant story of Chattanooga's demise and ascendancy has been powerfully constructed by a network of major institutional agencies – among them the local business-foundation alliance that crafted it and earned financial support, the news media that propagated it, the political leaders and civic directors who gained various forms of capital (votes, funding, influence) from supporting it and a presidential administration that legitimated it. That Cinderella story is so enmeshed in the officially approved record, popular media reports and political mythos that it is not easily shaken. Indeed, as we will show in a succeeding section, it is so powerful that it has succeeded in bringing continuity to the renewal of local efforts to once again draw political support and media attention in its replay in the fashioning of another phase of citizen mobilization behind a further rebranding of this city as "Chattanooga Green" (Chattanooga Green webpage, 2009) in a new business-governmental campaign beginning in 2005.

Informed by the strong sustainable city framework underpinning Part I of this work, we will seek to demonstrate how, far from advancing sustainable development, the Chattanooga example is really a mirage or chameleon that both conceals its distance from any strong version of sustainability and cleverly adapts to changing fads and fashions like a chameleon without significantly departing from its more deep-seated objective of balancing corporate-led economic growth and downtown revitalization with a measure of environmental responsibility and smart growth reforms.

Sustainability, Chattanooga-style

In Chattanooga's transit from "Environmental City" to "Sustainable City" brand, business and civic leaders were abetted in that smooth and unruffled transition by the "essentially contested" (Gallie, 1956; Connolly, 1983) nature of sustainability and its politics. Sustainability, like other political concepts (e.g., freedom, equality, the good society), involves political controversy, linguistic ambiguity and internal complexity, and an appraisal of existing conditions and their distance from the valued goal to be achieved. The city's major players who shaped its social, political and economic future portrayed sustainability in a variety of casual ways to answer the question, "What is sustainability?" For example, when David Crockett, Chattanooga city council member and director of the Chattanooga Institute, was asked what he meant by sustainability, he responded, "I'm not always quite sure what sustainability is, but I know it when I see it and I know what it's not" (Gibson 1998: 27).

This vagueness of meaning and understanding no doubt flows in part from the qualities that make sustainability an "essentially contested" concept. In the case of leading Chattanoogans, it derives from deeper reasons that have to do with how language can obfuscate and how ideology can shape politics and the social construction of reality. A close reading of the emergence of sustainability language into the official discourse of the Chattanooga economic renewal campaign reveals the extent to which the meaning of sustainability by this network of business, nonprofit and civic elites proved to be a solution to a political problem that stemmed from the instability of the initial branding of Chattanooga as the "Environmental City." Gibson takes note of the fact that, on the one hand, branding Chattanooga as the "Environmental City" mobilized both external and internal praise for the city in overcoming its legacy of environmental problems generated by its smokestack economy based upon polluting manufacturing industries. On the other hand, she notes, the marketing of the image of "Environmental Chattanooga" mobilized keen resistance from those corporate elites linked to manufacturing, who feared that the cost burdens of meeting stringent environmental standards would prove weighty and felt it would send a negative message to prospective businesses that might otherwise consider locating in Chattanooga and the region. On her reading, the banner of sustainable development, which was hoisted by the business-nonprofit-governmental coalition beginning in 1994 onward, proved to be a solution to this tension and division among Chattanooga business elites. This was precisely because its vagueness and ambiguity allowed the new marketing campaign to frame Chattanooga as the "Sustainable City" and retroactively recast its prior framing of the city as the "Environmental City" in a manner that permitted the relentless pursuit of economic growth (in terms of profits, jobs, income, new businesses and population gains) while muting the cost factors (like zero emissions, environmental remediation, toxic cleanup and the general corporate internalization of environmental damage).

As generally defined in public forums, the significant demands and economic implications of sustainability's stronger requirements and standards could be removed from serious consideration and active debate by interpreting sustainability as a process that, while incorporating economic, equity and environmental components, was never ending and would take forever. It is not for nothing that, as Portney (2003: 217) has observed, among the two dozen cities most often listed as leading sustainable cities in the United States Chattanooga has the distinction of being the solitary example that neither began its sustainability initiative with a set of indicators nor generated such indicators later to assess the progress toward achieving urban sustainability. While our book's strong sustainability approach has its own critique of indicators as yardsticks for measuring sustainability, arguing for a Sustainable

Area Budget as an alternative, Mark Roseland makes a germane point in saying that the virtue of sustainability indicators is that they assure that "what gets measured tends to get done," adding "if you can't recognize success, you can't reward it" and "if you can't recognize failure you can't learn from it" (Roseland, 1998: 203). Characterizing sustainability as an infinity point, as Chattanooga leaders did, absolves its practitioners of the responsibility of measuring progress and permits them to declare any seemingly green activity or project as a contribution toward the unreachable endpoint.

In many respects, the centerpiece of the Chattanooga strategy was the Chattanooga process – a model of visioning that was supposedly crafted to ensure wide citizen participation in working toward sustainability. Like the definition and strategy informing this work, the Chattanooga process underlined the local and participatory elements of generating community sustainability. As practiced in Chattanooga and proselytized in other cites by the Chattanooga Institute, local democratic participation tended to become more and more a mirage as the decision-making process moved from broad visioning exercises and general goal setting to concrete policymaking over priorities, methods of implementation and financing levels and sources. Given the powerful and influential roles played by the senior partners in the constellation of players involved in Chattanooga economic renaissance (i.e., the local chamber of commerce, the business community, the Lyndhurst Foundation and key public officials), the Chattanooga process came to resemble a pyramid of power where public feedback at the lowest visioning level gave way to the involvement of fewer and fewer players as alternatives and choices were discussed and decided upon in smaller and increasingly closed quarters. One mechanism used by the Chattanooga Chamber of Commerce and other organizations like the RiverValley Partners and the Lyndhurst Foundation for restricting broader citizen involvement was to price out prospective interlocutors by charging comparatively large sums ($500 and up) for admission to these supposedly public events. In addition, limitations on critical planning decisions of great import were often delegated to task forces and committees whose memberships were recruited from the "usual suspects" representing narrow economic interests (although token members from minorities and more vocal environmental groups were sometimes included).

The role of local government deserves comment, in this respect. During the implementation of the Chattanooga revitalization program, the local governmental process underwent a transformation from a political structure with a weak mayor-city manager system to a putatively strong mayor–city council system where council elections were district ones, not at-large seats. The expected impact was twofold: a strengthening of the office of mayor and the expansion of representation among minorities. To some extent, both anticipated outcomes were realized.

Despite these increases in mayoral influence and minority council representation, Chattanooga's pathway to downtown rebirth and sustainable economic development continued to be a guided by a business-nonprofit-governmental alliance where the corporate players were in the driver's seat and elected officials remained junior partners. Given Chattanooga's politically conservative nature, it is not surprising that in James Frierson's terms, "local government has actually been a junior partner by choice" (Frierson, 2003), making the local political institutions, in Parr's characterization, as "the caboose on the civic train" (Parr, 1998). In many respects, the political alliance structure and goals of Chattanooga's renaissance bear all of the earmarks of Clarence Stone's "development regime" (Stone, 1989, 1993; and Filion, Hoernig, Bunting and Sands, 2004) or Molotch and Logan's "growth machine" (Molotch, 1976; Logan and Molotch 1987).

Sustainability, Chattanooga-style, is best characterized as what it is and has always been: a large helping of downtown revitalization plus a modest measure of enforced environmental remediation plus a dash of smart growth. Much if not most of the substance of the Chattanooga strategy has been directed toward revivifying the downtown through high-profile public projects like the Tennessee Aquarium, the Riverwalk, the IMAX theater and the Walnut Street Bridge, which may be generously typified as "showcase sustainability," though sustainability in even its weakest sense is a stretch. Moreover, the prime beneficiaries of this public-private partnership in urban redevelopment have been many of those prominent partners in this enterprise. As Naylor (2003: 40) has acknowledged:

> When the "good old boys" decided to revitalize their city, they were partly motivated by personal gain. [Lyndhurst Foundation director] Jack Lupton's many real estate assets downtown have certainly benefited from the turn-around…[former mayor] Corker is Chattanooga's biggest commercial landlord; [and prior mayor] Kinsey is principal of the major real estate development firm that will soon begin work on several parcels in conjunction with the 21st Century Waterfront Plans.

Those who have other more personal, family and cultural stakes in the downtown, as we shall argue, have more often found themselves displaced from traditional neighborhoods and seen their interests marginalized in the processes of urban renewal and gentrification through such "Showcase Sustainability."

Chattanooga Creek and the manufacturing legacy of environmental racism

For much of Chattanooga's campaign to revitalize its downtown and rebrand its negative image constructed from its polluting manufacturing base, Chattanooga Creek symbolized both the challenge of its heavy industry legacy for this reimaging process and a test of its sincerity in repairing a significant environmental sore spot standing in the way of any refurbished brand. It was also a bellwether of how central to Chattanooga's environmental and alleged sustainability efforts was the pursuit of social equity – the third leg of sustainability tripod.

During the industrial heyday of Chattanooga, this tributary of the Tennessee River was "a veritable sewage canal for exploitive and polluting industries operating in a primarily African-American section of Chattanooga" (Bartling and Ferris, 1997: 3). Bordering this, the most polluted waterway in the southeast, were some 5,000 inhabitants (98 percent of whom were African-American) living in two residential tracts known as Piney Woods and Alton Park. For nearly seven decades, the land along Chattanooga Creek was the site for nearly all of Chattanooga's heavy industries (Rogge, 1998: 111). The historic neglect of this predominantly black residential section of Chattanooga mobilized serious concerns by its inhabitants about community health risks from the toxic chemicals dumped into the creek. It prompted local citizens, neighborhood groups and their indigenous leadership to press for quick remediation of this persisting environmental and health danger. After 30 years of community agitation to improve the living conditions in these two neighborhoods, the Chattanooga economic-environmental campaign led by economic and political elites triggered the proliferation of local community groups – including the Alton/Piney Woods Community Coalition, Stop Toxic Pollution (STOP) and Turning Point – to tackle this problem anew and determine how deeply, if at all, class and race biases were affecting this urban and economic renaissance campaign.

As these advocacy groups gained the attention of state and federal health agencies to investigate the public health hazards associated with the toxic industrial wastes in the Chattanooga Creek, these health assessments provided growing evidence of the risks this contaminated waterway posed to area citizens. As this evidence mounted and was amassed in a neighborhood data repository in the Alton Park Health Center, civic leaders connected with the Chattanooga revitalization program demonstrated little inclination to address this serious community problem and were slow to react to increasing alarm and impatience. By comparison to projects for downtown redevelopment and even air pollution remediation, the toxic cleanup of Chattanooga Creek was set off to the side and not faced with the same measure of resources, commitment and urgent attention. Such tardiness or inaction helped to fuel anger, suspicion and distrust among Alton Park-Piney Woods African-American residents, even toward nonresident environmental researchers carrying out an EPA environmental equity grant. As one black community leader voiced her anger and frustration, "We no longer need anybody that doesn't eat, sleep, and live in this community to come in with their pity and tell us what our community needs" (Henderson, 1995: E1, quoted in Rogge, 1998: 114). In contrast to the collaborative and at least formally participatory tactics embraced using the Chattanooga visioning process, leading neighborhood organizations who organized around this environmental and health issue were forced to engage in more confrontational methods to translate their seeming "private troubles" into "public issues" (see Rogge, 1998: 116; and Rogge, Davis, Maddox and Jackson, 2005: 40).

The cleanup of Chattanooga Creek was not addressed swiftly by town leaders, nor has it been completely remediated a decade or so since it began. Motivated by studies that pointed to the correlation between minority status and toxic risk and stunned by charges of environmental racism (Stockwell and Sorenson, 1994; Gibson, 1998: 56 and 87), Chattanooga's political and economic elites were successful in passing on the problem to the EPA and the Superfund beginning in 1994 when Chattanooga Creek was put on the national priority list. In the first phase of the cleanup in 1997–98, the EPA spent $12 million to remove approximately 25,000 cubic yards of coal tar residue from polyaromatic hydrocarbons that had been dumped into the creek by steel mills. This initial phase still left high levels present in the creek water, which poured into the Alton Park–Piney Woods communities in 2003 by major flooding from the worst storm in three decades (Rogge et al., 2005: 43). A second phase of the cleanup operations began in 2005 at a cost of $24 million. It involved additional remediation of the creek's underground surface and the installation of an Aquablok clay liner to prevent underground contamination from polluting the creek's flowing water (completed in 2008). As of 2009, reports have surfaced that the clay liner has been breached, allowing underground contaminants to leach into the water. Now, city officials, former industry polluters and federal agencies are squabbling over who should be held legally responsible for a long-term solution to toxic chemicals jeopardizing the temporary clay liner fix. Sadly, new polluting sources continue to plague Chattanooga Creek and the long-time neighborhood leader of STOP, Milton Jackson, stepped down from the organization, prompting speculation that STOP would shortly disband (Putman, 2009).

Chattanooga Neighborhood Enterprise and affordable housing

In a city where class, status and income levels were affected by Chattanooga's deindustrialization experience and where racism had its imprint upon housing patterns and availability, the issue of housing affordability was an obvious one that would have to be addressed in any economic and downtown turnaround. Beginning in the mid-eighties, Chattanooga Neighborhood Enterprise (CNE) became a key component of the Chattanooga revitalization initiative in its effort to respond to problems in rundown neighborhoods populated by substandard housing. As Chattanooga's renewal program was increasingly marketed as a sustainability enterprise, this public-private housing campaign functioned as a central agency in fortifying the social justice/social equity leg of the sustainability tripod.

Founded in December 1986, CNE grew out of the city's visioning process and a housing study undertaken by the Enterprise Foundation. Its latest, revised mission is "to improve the lives of low to moderate income Chattanoogans by building strong, sustainable neighborhood, and by fostering home ownership" (CNE website). In its early years, it was granted seed money by the Lyndhurst Foundation and an assemblage of banks, wealthy donors and corporations made a commitment to financially support its efforts over the long term. Its board of directors reads like the central committee of the larger Chattanooga economic/environmental/"sustainability" enterprise, suggesting how integral safe, fit and affordable housing and healthy neighborhoods are to larger efforts. Moreover, its list of accomplishments seems no less impressive. Over the course of its nearly quarter-century history, it has generated nearly 3,500 homeowner opportunities, fostered rehabilitation of 2,800 homes and utilized the services of the Community Development Financial Institution and Chattanooga Opportunity Fund to make available over $6 million to small and minority businesses and other local businesses located in lower income areas (CNE website). Recognizing revitalized neighborhoods buoyed by affordable housing as one measure of a sustainable city, it began in 1995 to scale up its activities to encompass sustainable neighborhood development.

Without diminishing the positive impacts this nonprofit housing and neighborhood enterprise has had on the individual lives of low-to-moderate citizens in formerly blighted areas of the city, it is legitimate to raise the issue of how to assess the successes and achievements of CNE's programs relative to the scope of the problem and the demographics of its patrons in the absence of meaningful yardsticks or indicators. Because Sustainable Chattanooga has eschewed sustainability-oriented indicators, CNE's record is replete with aggregated statistics and anecdotal evidence but lacking in specific measures of progress over time to assure, as Roseland has put it, "what gets measured tends to get done" (Roseland, 1998: 203).

As a result, anecdotal evidence raising serious questions about its achievements have been cited by critics of this affordable housing and healthy neighborhood program to cast doubt on its success in tackling a persisting social problem in Chattanooga's midst (Bailes and Mesh, 2005; and Rogge et al., 2005). For one thing, some critics point to the way that urban renewal projects targeting formerly blighted areas in and around the downtown have led to rising property values that have priced former residents out of the now upscale housing market in these areas; others have worried over whether efforts like the Hope VI public housing projects begun in 2003 will advantage only those with higher income, displacing many of those former residents who cannot afford the expense of single-family dwellings. Supporters of Chattanooga's housing programs administered by CNE and other city agencies argue to the contrary, stating like Frierson that "Alton Park is now at the epicenter of low-income housing development that has torn down large public housing buildings and replaced them with single- and multi-family housing" so "no one can say that the 'housing stock' for African-Americans has not improved" (Frierson, 2003).

In his case study of Chattanooga's ambitious urban renaissance initiative, Lancaster (2004: 198–199) fundamentally questions the degree to which CNE made a serious dent in the affordable housing problem in the predominantly African-American neighborhoods in the downtown and South Chattanooga. As he shows, using United States Census data from the critical 1995–2002 period, the escalating average prices of housing startups all but closed off minority residents of low- and moderate-incomes from taking advantage of new housing construction. He also draws data from other studies that demonstrate that in 2001–2002, African-Americans experienced mortgage rejection in 31.83 percent of applications made compared with only 18.86 percent among Caucasians; and while African-Americans reflect approximately 18.8 percent of Chattanooga's total population, they obtained only 4.21 percent of the conventional mortgages made there in 2002 (The Great Divide, 2003, cited by Lancaster, 2004: 199).

One may quibble over the relevance of the abovementioned statistics or take issue with their interpretation, but the central points are that: (1) the affordability problem for African-Americans is a significant problem for social policy; (2) the racial divide in Chattanooga continues to weigh heavily in limiting housing opportunities for this minority; and (3) the contribution of CNE is likely meager in comparison with the extent and seriousness of the problem. Without quantitative measures to track key dimensions of affordable housing (and in particular, African-American citizens' homeownership) and without yardsticks and targets to mark progress or regress on this key dimension of social equity, Chattanooga's venture in pursuing this critical component of sustainability will be rightfully questioned by those who champion the cause of social justice and urban sustainability.

CNE's subsequent history tends to confirm some of its critics' charges and concerns. Jean Lamb, Chattanooga's compliance officer, recently acknowledged that one of the side effects of CNE's projects was gentrification as older, dilapidated homes were razed and replaced with new ones selling for as much as $350,000. In her own words, "the whole idea of that area was to get mixed income housing and it's kind of picked up and gone on its own" (Carroll, 2009a: 1). Such gentrification of formerly older black neighborhoods stemmed in large part from a critical detour from its original mission and a foray into funding market-rate housing and rental property management in the first five years or so of the new century (Carroll, 2009a). By 2007, CNE's books were audited and poor management, casual financial practices, lax contractor hiring and oversight in violation of city bidding requirements were discovered (Flessner, 2007a). These instances of fiscal mismanagement prompted CNE to sell off rental units to private real-estate developers, who in turn, raised rents to market values, driving low-income inhabitants from their neighborhoods (Carroll, 2009b). Other consequences of the audit scandal followed: CNE's staff was cut from 56 to 19; city government cut the agency's funding from $2 million in fiscal year 2005–06 to $1.1 million in fiscal year 2007–08; and federal support likewise fell from $2.6 million to $1.4 million in the same period (Reynolds and Flessner, 2007; and Carroll, 2009b).

As it seeks to return to its original missions to increase homeownership of low- to moderate-income families and to assist homeowners in improving existing homes, CNE will have to deal with closer scrutiny from the city and greater oversight from the Lyndhurst Foundation, which by 2007 had funneled in excess of $23 million into CNE over the preceding two decades (Reynolds and Flessner, 2007; Putman, 2007 and Reynolds, 2007). In looking to the future, CNE now must work with a tarnished reputation, smaller staff and annual budget, more focused mission and the loss of its role as the affordable housing leader, and content itself with cooperating with other nonprofit groups as a partner (Reynolds, 2008).

CARTA and AVS – The transport mirage vanishes

Like many of the many other vaunted elements of the Chattanooga turnaround programs sporting a green hue, CARTA's germination of Advanced Vehicle Systems (AVS) and its production of electric buses were the talk of the town for its first nine years or so. With a $19.6 million grant from state and federal departments, Chattanooga was able to place its first order for 16 battery-powered buses, adding two more soon after. The incorporation and growth of electric buses in the city's downtown were ostensibly intended to reduce private automobile traffic in the city center and help alleviate Chattanooga's air pollution problem. As the "Environmental City" morphed into the "Sustainable City," its electric buses became a token of Chattanooga's programs and plans of turning green. As AVS president Rick Hitchcock put it, "the electric bus program has been the most visible and most successful element of this broad palate of sustainable ideas" (quoted in Porter, 2000: 4). The electric bus route also contributed to Chattanooga's downtown revitalization by linking tourists and downtown shoppers to businesses, hotels and parks in the central business district and had the virtue of being free of charge, since its operation was underwritten by the fees charged at parking structures built and operated by CARTA.

Early troubles with the new battery-powered entrants into Chattanooga's public transit system were addressed by the CARTA general manager Ron Sweeney, who relocated the bus maintenance area to the downtown to shorten the daily routes of this part of CARTA's fleet. Sweeney also solved an organizational structure issue stemming from maintenance workers who feared the physical risks of electrocution from high voltage and potential burns from handling the special flooded lead-acid batteries. He did so by retraining diesel mechanics and shifting them to the electric bus fleet after their apprehensions dissipated (Hess, 2005).

Problems of maintenance and reliability were overcome (apparently) and the Chattanooga electric bus fleet seemed to be popular among tourists and Chattanooga downtown shoppers. In a move to innovate electric-bus technology, AVS then moved onto building hybrid electric buses to extend the distance buses could travel. The fortunes of their builder, however, proved to be another matter. At a time when other cities around the nation became interested in purchasing electric buses or trolleys to enliven their central business district, AVS ran into mounting problems with those orders and sales. Over the first seven years, AVS built over 110 electric buses and expanded its market to Los Angeles and Sacramento, California; Galveston, Texas; Tempe, Arizona; Falls Church, Virginia; and Miami and Tampa, Florida. Though CARTA managed to keep its electric bus fleet running, the quality and reliability of AVS's public transit products came under assault, as orders to Tampa, Falls Church and Galveston were plagued with problems, leading to returns after persistent problems could not be overcome by AVS repair and maintenance workers, and Sacramento's five AVS electric buses were sold off after a short experience with nagging battery and engine troubles (Wexler, 2001).

The most visible emblem of Chattanooga's commitment to sustainability was dealt a severe blow in October 2003 when AVS filed for bankruptcy and closed its doors permanently, leaving CARTA's order for 11 next-generation hybrid-electric buses unfulfilled (PowerPulse, 2003). One assessment of the eleven-year-old electric bus manufacturer attributed AVS's demise to problems with the new gas-electric hybrid technology, which could not be overcome in time to save the company. An allied reason may also have been that AVS lacked research and development support that would have helped transcend early hybrid bugs (Hess, 2005). Its only organizational legacy was the emergence of a spinoff organization, the Electric Vehicle Institute, now named the Advanced Transportation Technology Institute, which explores electrical and other alternative fuels for future transport.

CARTA's sustainable transit program, however, was burdened by a deeper systemic flaw that was barely recognized and never addressed by leaders of the Chattanooga downtown revitalization initiative. While many sustainability advocates speak positively about electric buses as a good alternative for advancing public transit as part of a larger urban sustainability program, Roseland makes the critical point that the "concept of 'sustainable transportation' calls for a more holistic approach to community planning, policy, and investment… to fulfill mobility needs" (1998: 111). That is, unless sustainability-oriented transportation policy is integrated into a strong sustainability framework that includes green economic development, land-use planning, downtown design (including population density), and even agriculture and food policy elements, it is likely to succumb to isolated, incrementalist reforms that undercut its role and potential for contributing to urban sustainability. The sad saga of AVS and the continuing place of CARTA and its electric bus fleet within Chattanooga's largely smart growth policy frame demonstrate that Roseland's key insight has not yet been appropriated and institutionalized into community planning.

Chattanooga Green: Entrepreneurial "sustainability" redux?

Between 2000 and 2003, the visible signs and tokens of Sustainable Chattanooga disappeared. Emblematic of the fading of this city brand were a number of subtle and not so subtle changes that eroded the strength of Chattanooga's image. The portal website, Sustainable Chattanooga, started in 1994 by Chattanooga Community Link and funded initially by a Lyndhurst Foundation grant as a web-based forum for disseminating the Chattanooga way to other cities and towns, ceased being updated by the summer of 1997. The Chattanooga Institute, directed by former city councilmember David Crockett, experienced a meteoric rise through seed funding by Lupton's foundation, but eventually closed its doors after its failure to catalyze cross-state and regional support for Chatlanta, a Maglev train initiative. Chatlanta would have connected Chattanooga and Atlanta and given true regional focus to Chattanooga's endeavors to link commerce, culture and collaboration (Chattanooga Institute website, 2000). Even more telling, with the election of Jon Kinsey as mayor in 1997 and his expressed skepticism about the direction taken by Chattanooga, the local chamber of commerce backed away from the sustainability brand. Stung by criticisms of the sustainability label and its markets by elements of the area business community in the late nineties that "Sustainable Chattanooga" had negatively affected industrial recruitment, the chamber moved to align itself with Kinsey's promotion of a more "diversified" approach to urban economic development. That orientation was recentered on industrial recruitment and job creation by rechristening Chattanooga's business image and message through a new campaign organized around the slogans "Chattanooga Can Do" and "Tell the World" (Flessner and Sohn, 2008).

Like the capacity of chameleons to change colors and blend into the shifting natural landscape, Chattanooga found itself at another crossroads in a political debate over its ambiguous present and uncertain economic future. This was expressed in the 2005 mayoralty contest between Ann Coulter and Ron Littlefield. Both Coulter and Littlefield were heavily involved in Chattanooga's urban-economic-environmental campaigns of the 1990s, but Littlefield emerged as the winning candidate. He apparently won because he raised the most questions about the old sustainability campaign's economic impact and, as mayor, promised to restore a focus on old-style economic growth, industrial recruitment and job creation.

Under the new mayoral administration, the contours of Chattanooga's community development campaign took two interesting pathways. On the one hand, Mayor Littlefield honored his campaign pledge to renew focus on attracting new industry and jobs. He launched an aggressive industrial recruitment campaign that netted Chattanooga a $200 million Alstom nuclear-power turbine-manufacturing plant (Chattanooga Chamber of Commerce, 2007) and two years later, a $1 billion Volkswagen auto assembly plant (Chattanooga Chamber of Commerce, 2008; Volkswagen, 2009). Together, these new plants were expected to bring over 1,350 jobs to Chattanooga-Hamilton County (Hashe, 2009). The impact upon strategic sectors of Chattanooga and the wider region of these developments is sizable – something akin, as one observer characterized it, to "absorbing a small town over a seven-county region" Carroll, 2009b).

The other path opened up by Littlefield's election proved to be a somewhat surprising one: the remobilization of the earlier supposed sustainability vision now branded "Chattanooga Green" and the "greening of Chattanooga." In his first term, Littlefield launched a number of initiatives, some harkening back to the earlier "Environmental City"/"Sustainable City" campaign, while others promised renovations of the older brand. In the fall of 2007, the mayor appointed a twelve-member Green Committee to advise his office and the council on sustainability issues. The following April, that committee hosted a visioning session to engage community citizens on green issues and to offered recommendations for implementation. It issued an interim and then a final Climate Action Plan in 2008 and January 2009 respectively (Chattanooga Green Committee, 2008; 2009; Hightower, 2009a). Reflecting mounting evidence of a scientific consensus over the causes and potential consequences of global warming and growing popularity for green procedures and practices, the mayor signed Chattanooga onto the United States Conference of Mayors climate change agreement in the spring of 2008. This was followed in late 2009 with the establishment of an Office of Sustainability staffed by a director and two interns. The office's director is specifically responsible for setting priorities among the proposals formulated by 14 climate-plan action committees (Hightower, 2009b).

The prospects for Chattanooga's new green enterprise breaking out of its predecessor's smart growth/weak sustainability framework and becoming stronger and more holistic in relation to the urban sustainability rubric outlined in Chapter 3 are uncertain. For one thing, this new brand continues to inhabit the lower end of the sustainability scale comprising the Urban Sustainability rubric. Typifying this new community campaign as "green" tends to carry with it the biases and limitations associated with the green vs sustainability dichotomy previously delineated.

In addition, the various offices, committees and organizations involved in advancing this green agenda are led or populated by many of the same cast of characters who earlier were prominent movers and shakers in the Environmental City-cum-Sustainable City phases of the Chattanooga version 1.0. Karen Hundt, previously director of the Regional Planning Association Design Studio, is now an influential member of the Green Committee. The inaugural director of the Office of Sustainability is David Crockett, former councilmember and Chattanooga Institute director (Sohn, 2010). James Frierson is a major player in the chamber of commerce and presently directs the Advanced Transportation Technology Institute. J. Ed. Marston, a key economic booster of downtown revitalization, remains a central figure in the Chattanooga Chamber of Commerce.

Surely, a healthy dose of innovative thinking needs to be injected into Chattanooga Green 2.0 if it is to rise above its previous foundations and shake loose from the economic growth, racial biases and class interests informing its previous iterations. The fruits of Littlefield's industrial recruitment strategy will also challenge it.

Toward Chattanooga Sustainability 3.0 – Some Policy Recommendations

In this concluding section, we consider some policy recommendations that rise to the level of strategic interventions in the scope and substance of Chattanooga's sustainability enterprise.

Shift from smart growth and weak versions of green planning policy to strong urban sustainability. The entire trajectory of the Chattanooga's urban rebuilding program has been constrained by guiding assumptions informing the chamber-led growth coalition that has been in the driver's seat three decades or more. Generally pliant political leaders and representatives have served vigorous and clever business leadership organized around the local chamber of commerce in local government. This business-foundation-governmental alliance has hobbled efforts by citizen-action groups like the Environmental City Task Force and downtown social justice organizations to embrace any variety of sustainability that might weaken the elite coalition's grip on the city's agenda for social and environmental change. With the community assets and social capital that have built up over the years among Chattanooga elites and citizen activists, it is time for those resources to be marshaled behind the cause of strong urban sustainability projects and activities.

Bypass the sustainability indicators approach and develop the city-region on the basis of a Sustainable Area Budget. A cardinal criticism of Chattanooga's renaissance program, which we continually emphasize in this chapter, is its treatment of sustainability as an interminable road and its failure to utilize quantitative markers of progress on that road. Elsewhere in this volume, we have criticized sustainability indicators as poor instruments for achieving strong urban sustainability. Chattanooga would make a significant step toward embracing a strong version of urban sustainability if it were to encompass the city-region as the scale of sustainability activities and work out calculations to determine its Sustainable Area Budget (SAB). With this formula, it would be able to have a sustainability yardstick that would allow it to permit an array of sustainability tools, like the Sustainable City Game, multiple scenario building and, when available, the Sustainability Engine, to inaugurate a process leading to institutionalization of a *local, participatory, balance-seeking process* and other features comprising the essential elements of strong urban sustainability.

Move from Chattanooga's model of visioning and participation to the Sustainable City Game and multiple scenario building. As our critique of the Chattanooga process unfolded, we sought to demonstrate how a visioning process standing at its center opened out onto a hierarchy of effective decision making that became more and more restrictive as its moved upward in terms of stakeholders and increasingly closed to everyone but those whose wealth, status and power were drawn from elite sectors within the business and social communities. While it may be difficult to envisage and institutionalize – especially in a community where power and influence have congealed into informal circles of the upper members of the largely governing growth coalition – the Sustainable City Game can actually be played in a host of different formal and informal institutional settings. It can also foster modes of multiple scenario building, which can then lead to creative and collaborative policy solutions to social and ecological issues; help overcome entrenched economic interests; and have the virtue of involving (at the beginning and the end) more inclusive democratic participation within the community.

Fortify the third "E" in the sustainability tripod. If there is one strategic intervention in Chattanooga's political processes that would make a significant difference in its quest for sustainability, it is the greater and more inclusive involvement of minorities in the large issues that define and ground strong urban sustainability. Despite over $2 billion being invested in housing in the downtown, the homeownership rate of 55.7 percent in Chattanooga continues to lag behind the state of Tennessee as a whole (70 percent) and the nation (67 percent) (Putman and Reynolds, 2009). More open processes of decision making founded upon the model of the Sustainable City Game would not only help to overcome deep anger and suspicion from African-Americans and other minorities who suffer disproportionately from environmental and public health risks (such as were evident in the remediation of Chattanooga Creek). It would also draw upon the local knowledge and distinctive cultural assets of such communities to integrate and strengthen the social-equity leg of strong sustainable city-regions.

Work toward greater population density in the city core. As Chattanooga strives to increase the number of good-paying manufacturing and other jobs in the green economy, green economic policy must take into account the impact of such strategies and tactics for land-use policy. In particular, it must take into account the necessity of creating greater opportunities for reversing urban sprawl and attracting more and more Chattanoogans back to the downtown without marginalizing the homeless or disenfranchising minorities who had formerly lived in aesthetically blighted but often culturally lively downtown neighborhoods. Chattanooga and its surrounding towns and counties will be challenged to find creative ways of responding to the impact of Volkswagen's and Altsom's investments and growth of new jobs for the region without contributing to even more suburban growth and residential segmenting. As the most recent Chattanooga Region Report on Environment by the Chattanooga-based Ochs Center for Metropolitan Studies revealed, metropolitan Chattanooga's carbon footprint has continued to increase as a result of urban sprawl and population increases, while its air quality has worsened and its waterways have become more impaired (Tharp, 2008). The place to start is with serious rethinking of the possibilities of greater densification of Chattanooga's downtown. As the Ochs Center President David Eichenthal underlined, "Clearly there is a relationship between land-use, open space and transportation patterns that gets translated into air quality and even water quality in our region" (Sohn, 2008: 1).

Advance sustainability-oriented transportation policy by reducing automobility and integrating land-use policy and other sustainability-oriented policy reforms. As we noted earlier in this chapter, an innovative, sustainability-oriented transportation policy can help address the sprawl issue. By integrating transportation policy decisions with other elements of urban sustainability (including land-use planning, business growth incentives, downtown housing policies and tourism), effective programs can be created for translating downtown living, central city employment opportunities, public transit and bicycle commuting into a more holistic sustainability program that produces synergistic outcomes.

Fulfill the early dream of true regionalism. If, as we have argued, the basic scale of strong sustainability is the city-region, Chattanoogans will need to draw upon the wisdom of early pioneers of creative regionalism by returning to the insights of people like Lewis Mumford and the all-too-short-lived Regional Policy Association of America (RPAA). Urban sustainability programs with substance and direction like Vancouver have generated a long history of building regional planning organizations to address the encompassing nature of policy issues in water conservation, agricultural production and food security, energy efficiency, transportation policy and healthcare provision. Despite its metropolitan clout, Vancouver has come to realize that sustainability requires close cooperation and collaborative planning to avoid zero-sum policy outcomes. The Chattanooga initiative in renewal and revitalization, through its various iterations over the last forty years, has had the virtue of always coming around to worrying over and taking halting steps to initiate conversations about regional design. This dialogue has been abetted by the cooperative planning agencies it has created with nearby towns and counties to confront the regional context of those many issues in need of addressing beyond city borders. Those overtures and efforts have only been diminished by the underlying economic growth imperatives that have informed them. Perhaps Mayor Littlefield's latest call for renewed hard thinking and action in promoting regional planning and cooperation in the context of his green initiatives and economic recruitment strategy will provide a new occasion to get regionalism as an integral component of urban sustainability done right (Littlefield, 2008).

Rebalance the roles of business, government, and civil society. If these strategic interventions are to take place, the prevailing vectors of forces shaping the contours of Chattanooga's economic, environmental and social future must be squarely challenged through democratic processes and forms of collective action. Without effective power, these strategic interventions will at best inhabit the margins of public discourse or not occur at all. That means that politics is ineluctable and the developmental regime or growth coalition presently holding hegemonic power in Chattanooga will have to be surpassed. It also means that the differential role and influence among the institutions of business, government and civil society will need to be reinvented and rebalanced. The greater assertiveness of the mayor's office in the most recent iteration of sustainable themes, illustrated by the appointment of the Green Committee, the formulation of a local climate action plan, the establishment of the Office of Sustainability, and the hiring of the American Institute of Architects' Sustainable Design Assessment Team, are encouraging signs and possible steps forward (on the last see: Bailey, 2009; Boniface, 2009; and Hashe, 2009). But the absence of strong and vocal citizen action groups for sustainability – exemplified in the recent past by the Environmental City Task Force – to monitor governmental and business actions and to partly counterbalance heavy corporate influence is dismaying. One test of Chattanooga's progress toward strong urban sustainability will be its capacity to develop vibrant countervailing civic groups capable of steering public discourse, generating new stories and recovering the public commons that we as embodied creatures share with Earth and all its other creatures.

Chapter 9

SUSTAINABILITY COMES TO OKOTOKS, ALBERTA

Introduction

Nestled along the Sheep River Valley in farm country some 20 km south of Calgary, Alberta, the quaint Canadian town of Okotoks (pop. 11,664; 2004) presents itself as an inviting focus of investigation for anyone interested in the advancement of sustainability at the local, regional, national and even global levels. The winner of multiple awards, including the Federal Canadian Municipalities Sustainable Communities Award (2000), the Alberta Emerald Award for Government Institutions (1999 and 2002) and the International Dubai Award for Top 100 Practices (2000), its sustainable community program has attracted worldwide attention and growing emulation among small towns and communities in Canada and the United States. The web portal for the Town of Okotoks (website) and Sustainable Okotoks (website) convinces the reader with its sophistication and information-packed content that however committed it may be to retaining its small town environment built upon a village concept, it is a community on the move with exciting plans for enriching its heritage – and living within the bounds of its water supply's carrying capacity. The town motto, "Okotoks: Historic Past, Sustainable Future," serves as a warm homage to the past and a bold promise for the future.

This chapter offers an analysis of the state of the town sustainability program in Okotoks, Alberta. It seeks to measure the promise outlined in documents establishing a sustainable vision for Okotoks' future formulated in 1998 and later ratified in a series of public forums and local council meetings against the realities of their implementation and struggle with regional developments impinging upon them. It also seeks to explore the further development of Okotoks' sustainable initiatives in the face of contradictory trends and unexpected obstacles, some of which have derived from assuming regional pressures emanating from its nearest metropole, Calgary. In this effort to set its prevailing realities against its heralded potential, we close this chapter with a set of policy recommendations flowing from personal observations, interviewing and assessing the state of current projects comprising this ambitious town's local sustainability enterprise.

Present at the Creation: The Origins of Sustainable Okotoks

The official story of Sustainable Okotoks is the account of an innovative and farseeing municipal manager with a better idea and an in-house planning team with a superior vision of the future. It is also a tale of a small town at the crossroads. Comparatively, it is a narrative about the distinctive resources of Canadian political culture that are often found in such short supply by its friendly neighbor and vaunted superpower to its south.

The origins of Sustainable Okotoks can be traced to the appearance and shaping hand of Will Pearce, Okotoks' municipal manager, who joined the town administration in the mid-eighties and exerted growing influence over its direction and contours. Working within the confines and possibilities opened up by the town's need to update its General Development Plan and accompanying land-use bylaws, Pearce took the pressures flowing from significant population growth over the preceding five years and the opportunity presented by the town's regularized planning schedule to fashion a visioning and planning process that would balance Okotoks' administrative initiatives and citizen participation into a program that would become the model for other small towns in Canada (Limbird, 2003; Parkinson and Roseland, 2002; and Pearce, 2002).

Starting with the Okotoks' vision document (inherited from the previous Municipal Council), the municipal manager and his planning group engaged in a regular dialogue with council members after city council meetings as the vision of Sustainable Okotoks unfolded. As a result, council members became well-versed in what the vision meant in concrete terms and what policy implications would flow from it. In these informal sessions, options were laid out, the pluses and minuses of each proposed program were examined, and eventually a consensus was reached among the innovative municipal manager, a progressive council and the town mayor.

Two other key facets of this early planning effort were notable: a public education campaign mounted in the weekly newspaper, *Western Wheel*, to inform citizens of the council's planning vision; and periodic polling of citizens' attitudes toward the town and its future (Sustainable Okotoks, Public Participatory Process, 2005). The community education effort included newspaper articles highlighting the specific components of the sustainability campaign, supplemented by inserts in utility bills and other venues to "test the waters" of community sentiment toward the Sustainable Okotoks vision (Pearce, 2005). Beginning in 1997–8, the community surveys evidenced widespread public concern for the potential loss of Okotoks' small-town character. They also revealed strong support for taking a new course in municipal policies that embraced controlled growth. Along with public hearings and ongoing town council meetings, the community surveys became an integral and regularized element of Sustainable Okotoks, serving as a tool for gauging public sentiment on planned programs and policies convergent with its vision and general goals. Despite the fact that the initial process was formulated and guided by a four-member staff within the town planning offices, this small in-house team eschewed the idea of instituting a municipal task force. Instead, the team decided in favor of launching an intensive and open citizen participation process in order to avoid tendencies toward top-down technocratic decision making (Pearce, 2002). (Given the strength and prerogatives of the Canadian civil service and greater trust of Canadians in government and business, this decision gave a strong boost to grassroots participation in the evolving community sustainability programs.)

One of the most critical decisions affirmed by the wider community was the choice to establish a "growth envelope" around Okotoks to limit town population growth to 25,000 to 30,000 inhabitants. The population cap on this growth boundary was based on calculating the maximum carrying capacity of its water supply from the Sheep River. This limit was chosen after reviewing various other means of imposing population and growth caps – including Wackernagel and Rees's "ecological footprint" analysis (1995). With this means of checking town growth integrated into the Municipal Development Plan (MDP) and its land-use bylaws, planning mechanisms were put in place that permitted the town council and administration to achieve control over the overall direction of future development as well as shape the location of industrial, commercial and residential sites and impose limits on the size of infrastructure allowed in those new development zones.

The MDP, titled "The Legacy Plan" (Town of Okotoks, MDP, 1998), outlines in great detail the many dimensions constituting the town sustainability design including downtown development, transportation, parks and pathways, urban forests, shoreline and escarpment areas, commercial and industrial development, community facilities, schools, financial and utility programs, and the urban/rural interface. For a small town of 11,000 inhabitants at its formulation, the MDP exhibits great sophistication and promise as a framework for community sustainability. It also sets targets to evaluate the success of the plan and to assay whether it is being properly implemented. Among the list of targets specified, two stand out: (1) increasing the town's industrial and commercial assessment base to 22 percent of the total tax assessment at build-out to ensure, among other things, that the town's sustainability program can be adequately financed over the long term

(Town of Okotoks, MDP, 1998: 48 and 61); and (2) embracing a goal of 30 percent nontraditional dwelling units to foster diversification of housing options that would allow single-parent families, unmarried individuals, senior citizens, and lower income families and wage earners to find places in the sustainable community of 2020 and beyond (ibid.: 51). In partnership with the MDP, the Town of Okotoks' land-use bylaws served not only to fill out the details and implications of the municipal development blueprint; it also functioned as a set of land-use regulations and building restrictions to strengthen the hands of town planners and administrators. The latter would assure that the growth envelope established by the cap on population would not be compromised by mounting population growth and political and commercial pressures to overstep the limits set by that cap and the larger program's commitment to local sustainability (see also Limbird, 2003; Tools of Change, 2000; and Pearce, 2002).

In addition, recognizing that town plans would need to be supplemented with regional planning, Will Pearce and other administrators worked to consummate an intermunicipal planning agreement with surrounding towns. In 1998, Okotoks joined the Municipal District of Foothills in adopting an intermunicipal development plan (IDP) founded upon principles underpinning the Sustainable Okotoks model. Its intent was to "establish the build-out Town limits through mutually endorsed statutory plans, and provide for a mutual development referral process between the two municipalities" (Limbird, 2003: 5). No less important, the IDP was negotiated to encourage Foothills to cluster rural development around Okotoks in order to create a "privately owned green belt" that would prevent unwanted commercial and residential development from leapfrogging the growth boundary into other surrounding areas (Pearce, 2005). This IDP was complemented by the consummation of a tricommunity watershed initiative in March 2002 that brought the towns of Okotoks, Black Diamond and Turner Valley into the Canadian Community Monitoring Network. The purpose of this initiative was to assist the three rural municipalities in the task of promoting water sustainability by better coordinating the monitoring and management of their shared water resources (Lynch, 2005).

Finally, the Town of Okotoks developed a partnership with the University of Calgary's environmental design program headed by Mary-Ellen Tyler to undertake a three-year design demonstration project (1999–2002) to investigate ways to implement the Sustainable Okotoks strategy (Faculty of Environmental Design, n.d.; and Ballance, 2001). In collaboration with TransAlta (Canada's largest power generator) and Prominence Realty, the faculty and student research team took the developer's 142-acre site (called Tucker land) and visualized a mixed urban-residential design based on sustainability precepts. According to Will Pearce, although the developer never aggressively pursued the design concept, the visualization exercise was a "tremendous learning experience" (Pearce, 2005) as a number of the design components were transferred to other Sustainable Okotoks development projects. In addition, these design principles were subsequently integrated into designs being developed in Cranbrook, the community where Pearce presently serves as city administrator (Pearce, 2005).

With these elements in place or on the drawing boards, the Town of Okotoks began its journey toward community sustainability.

Phase 1: Native Resources, Promising Developments and Unresolved Problems

The first-time visitor to Okotoks is immediately impressed with the significant residential housing development apparent on the outskirts of its historic downtown and the surrounding commercial sprawl of this small Alberta community, whose population now exceeds 17,000. Much of the housing stock is new or built within the last five years; architecturally, much of it is beautiful, energy efficient and largely free of the row upon row of tacky designs so characteristic of housing developments in small-town America. Even apartment complexes, painted in rich primary colors, are aesthetically pleasing. The surrounding commercial landscape on Southridge Drive (its southern business extension), on the other hand, conforms to the grab bag business development that James Howard Kunstler so mercilessly criticized in his book, *The Geography of Nowhere* (1994).

That prodigious commercial and residential development along Southridge Drive has not been kind to Okotoks' historic center. Ironically, the traditional downtown business and government district has been largely decentered by explosive growth to its south and northeast. Given its proximity to the meandering Sheep River, Okotoks' downtown has experienced periodic flooding from the river's swelling from heavy rains. During June 2005, Okotoks suffered three floods along the downtown's Elizabeth Street and Riverside Drive and South Railway Street, which caused damage to its main underground sewer line below the river. As a result, fairly extensive – and expensive – repairs had to be undertaken. Meanwhile, a major upgrade of the Okotoks wastewater treatment facilities at the end of the downtown area had to take place (A. Berry, 2005).

One of the bright spots of the Okotoks sustainability initiatives is its progress in advancing a routinely underdeveloped component of overall sustainability; i.e., cultural sustainability. The Cultural Station, located in the former Okotoks' railway station, is one manifestation of such sustainability. Even though the

Figure 9.1. Okotoks map

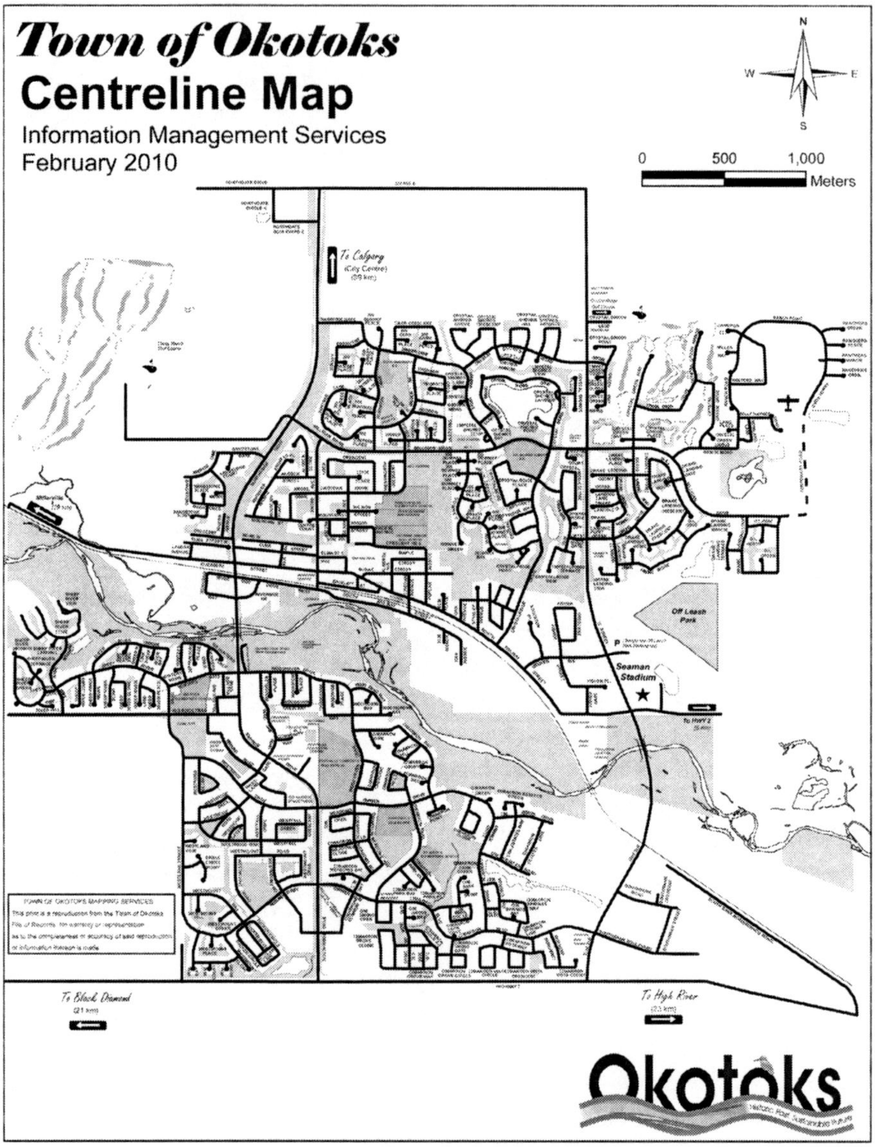

Source: Town of Okotoks Planning Department

train does not stop here anymore, local artists and crafts people show their cultural wares in viewings publicized through various outlets to the entire community. During early August 2005, the Cultural Station promoted the Art Walk, an unguided tour of downtown businesses where paintings, sculptures and other artwork were on display. In addition to acquainting tourists and local townsfolk with the talent of local cultural arts and crafts, the tour drew them to local businesses in the downtown core.

Historic downtown also features the Okotoks Heritage House, a two-story formerly private residence. This cultural site houses the town's historic archives, permanent and revolving exhibits and displays, and a staff of one. At the time of the field study, the heritage and preservationist museum offered an impressive display of photos and text reviewing the impact of the three June 2005 floods along the banks of the Sheep River, affecting Okotoks and other communities along its path, like High Hope. As one panel revealed, the Sheep River's banks had been breached five times in the twentieth century (1902, 1915, 1932, 1963 and 1995). The latest floods engulfed the downtown corridor and, for a time, traffic was closed off from Northridge Drive to the Southridge Drive commercial domain. Despite this calamity, which resulted in over $5 million in damage (not including commercial and residential damage) and the legacy of past flooding, no long-term solution has surfaced or is planned.

Indeed, the MDP only briefly mentions the problem of periodic downtown flooding, and merely refers to the Flood Mitigation Alternatives report released in February 1988 but evidently not acted upon (Town of Okotoks, MDP, 1998: 37). Instead, most public work has centered on clearing and restoring the river's channel, replacing sewage lines that were ruptured by the flood and cleaning up the mud and other detritus swept up onto public and commercial properties in the downtown district. The visual and informational exhibits underline the fact that the waterway both sustains the lives of those who inhabit its valley and periodically threatens their economic well-being, public health and safety.

The second display of relevance to this fieldwork was titled "Okotoks: Gateway to the Oilfield, 1913–1952." Through old photographs, memorabilia, records and other textual material, the exhibit highlighted the central place of Okotoks in the discovery and exploitation of oil reserves in the Turner Valley to the town's west. By happenstance and opportunism, the Turner Valley oil reservoir turned Okotoks into a crucial trade and transportation hub – one that brought considerable economic progress and a sizable stratum of local business people, wealthy and influential scions of the town (and even region). As the display legends pointed out, when these nearby fields began to peter out, more extensive (and therefore more profitable) oil stores were discovered in Leduc and Redwater; these fields overtook Turner Valley's diminishing oil production, erasing virtually all traces of its positive impact upon Okotoks' economy.

One of the most heartening developments in the initial phase of Sustainable Okotoks' planning process was the town's recycling depot. The facility is often a beehive of activity made up of individuals and families busy unloading, separating and depositing plastics, glass, paper (mostly newspaper) and sometimes cardboard. Simultaneously, town trucks frequently deliver the collections of newspaper, glass bottles and cardboard to the recycling site, which go into recycling machines (a newspaper bailer, a glass crusher and a cardboard compactor) and are transformed into shattered glass, four-foot square bales (in the case of cardboard) or into large bins of shredded paper. The paper is eventually sent to nearby Calgary or Edmonton, where it is made into insulation or the paper cover on wallboard; the cardboard is shipped to Tacoma, Washington, where it is recycled into various uses including being made into a wood substitute; and the glass is trucked to a glass recycling plant in Vancouver, British Columbia, where it serves a variety of purposes (McDonald, 2005).

Darryl McDonald is the recycling center's manager. According to him, in 2004, residents recycled over 1,500 tons (120 tons more than the previous year) of paper products, plastics and glass that would otherwise have been buried in the landfill. Since the town does not pick up recyclables at residential or commercial curbs, the recycling depot depends upon the three R's ("reduce, reuse, recycle") ethic of its citizens to individually separate out recyclables and personally transport them to the facility. In the case of plastic bottles, the recycling center requires conservation-minded residents to remove labels and clean them before depositing them in appropriate canisters on site. By doing so, these collective efforts save taxpayers (i.e., themselves) $25,000 to $30,000 annually that it would otherwise cost for the depot to hire another worker to do this job (Town of Okotoks, 2005).

Policy regulations and ongoing practice in the town affirm that Okotoks is serious about recycling. City garbage collection workers are bound by local regulations to pick up no more than three bags of garbage per week from each residence. The public information message justifying this ordinance underlines the point that if citizens are having a hard time living within this limit, they should look to recycling as a solution.

Since ground was broken, Drake Landing, an enclosed solar-energy neighborhood community, has been touted as Canada's first all-encompassing solar-energy housing development. It is located a scant 4 kilometers north of Elizabeth Street on 32nd Street. This solar community stands out for its primary and secondary solar-energy technologies, which are designed into individual homes and in a coordinated way (in which thermal energy is used) to tie renewable-energy production of individual houses into a whole (Drake

Figure 9.2. Okotoks population by age (2006)

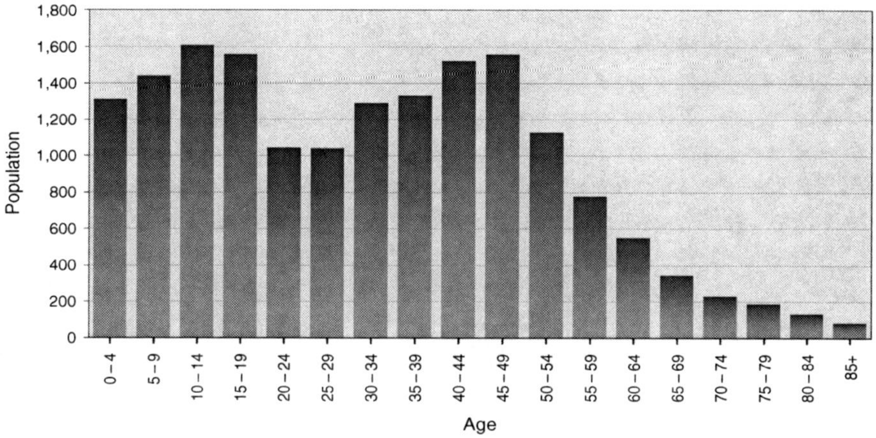

Source: Statistics Canada, 2007

Landing Solar Community, n.d.). But Drake Landing remains woefully lacking in its suburban layout of row upon row of houses, with undistinguished two-story designs. This suggests little forethought about land-use innovations, novel building design or mixed-use assemblage.

Given the extensive land surrounding Okotoks historically given over to farming, the town is both blessed and cursed with a lot of flat land on which to expand. As a result, there seems to be little incentive to build housing or mixed-use commercial buildings or communities contiguous with the downtown area. The most disturbing thing about the Drake Landing development is that it is located in the middle of a number of other conventional housing developments (e.g., Crystal Ridge and Crystal Green) several miles northeast of Okotoks' downtown center. Large housing tracts are increasingly taking their entropic course and being clustered into satellites (sometimes we call them suburbs) of the old town center. Where residential growth goes, it has been an American (and Canadian) historical propensity for commercial development and large churches to follow shortly after. It will take powerful community unity and a stringent, legally unchallengeable zoning process as well as other well-conceived local regulations to prevent this unsustainable land-use pattern from recurring.

The spread of commercial development around Okotoks – not only to the south, with residential development west of the historical downtown, but also residential clustering north and northeast – points to the absence of a strong and visible transportation component in the town's sustainable development strategy. Small as it is – or rather has been – Okotoks has not seen the need for any kind of public transit to curb (or at least complement) the private automobile. Yet as Okotoks' population spreads out, some kind of public bus service tying the town together will be needed – that is, if the town fathers and mothers have any hope of reducing the increasing automobile traffic that will snarl its streets and roads, and if they wish to keep the historic downtown a hub of thriving business activity. There are no guarantees that public transportation will be popular or that it will be patronized, primarily because the residential areas that would tie into the downtown are predominantly populated with new, young and professionally upwardly mobile individuals and families (precisely the constituency least likely to be lured to bus riding).

Okotoks' MDP is premised on the graying of its population over the next decade and a half. At the same time, its town planners are depending upon its quality of life and small-town allure to attract young knowledge workers and technology professionals to their vision of an autonomous community and increasingly high-tech economy. In this light, two key players in the ultimate outcome of this energetic and earnest local sustainability project will be the youthful demographic population group being lured to Okotoks, and the mature and aging adult demographic group present at the inception of Sustainable Okotoks. If the younger constituency is tempted to make Okotoks over into a modern city with urban attractions and amusements, it will negate the project's overriding commitment to a "no-growth" policy once the carrying capacity of its water supply is reached (at around 25,000–30,000 residents). If the commitment of the aging adult population flags – or is overwhelmed by the influx of young professionals seeking an urban scene and entertainment hub that approximates a mini-Calgary – all bets are off on the realization of this local sustainability vision for Okotoks.

A variety of sources unearthed during field research pointed to growing ambivalence, even discernible disenchantment, over the long-term future of Sustainable Okotoks. Uncertainty over the course of this sustainability policy initiative and the prospects for realizing its ultimate objectives was influenced and chastened by recent developments in High River, Okotoks' perennial rival in the region. Historically, High River has been the area's more prominent economic force, benefiting from a larger and more durable infrastructure than Okotoks'. High River's more mature political and civic base was recently evidenced in its rapid and sophisticated response to the June 2005 flood that washed over the Sheep River's flood plain.

High River's problem, however, according to one source (Green, 2005), is that in recent years its political direction has come under the increasing sway of prodevelopment interests that have managed to strike deals with town officials. It has been so successful that this progrowth alliance now threatens to turn High River into what sociologists Harvey Molotch and John Logan (1987) have termed a "growth machine" by unseating controlled growth incumbents in the town's planning and administrative offices.

Citizen activists place some hope in the existence of Healthy Okotoks (a small citizens group whose local actions are based on a medical model of community livelihood) and the emergence of land-trust organizations like the Sheep River Land Trust to blunt efforts at converting increasingly high agricultural land values into land speculator dreams of expanding residential tracts and commercial development on what was formerly farmland. Still, fears are mounting among controlled-growth loyalists that such groups alone may not be enough to stem the tide of galloping housing and business development. Such growth is overtaking High River and is rampant northeast and south of the Okotoks' downtown corridor.

A much more sanguine assessment of current trends and the town's future prospects was voiced by Alyssa Berry, the Town of Okotoks' corporate communications specialist. Okotoks' original sustainability vision was carefully advanced by Municipal Manager Will Pearce in 1997–8 and was institutionalized into its municipal development plan, its intermunicipal plan and its business plans in subsequent years. The progress of this vision, from Ms Berry's perspective, is on track and the goals of the master plan and the limits instilled in this document – to restrict further population growth above 25,000 to 30,000 inhabitants – are well understood by local citizens, land developers and construction company owners.

Since the town council and mayor ratified the plan in 1998, the Town of Okotoks has surveyed residents about their feelings and attitudes toward the means and goals of the program. With each new survey, strong public support for the local government's policies comprising Sustainable Okotoks has held firm or increased almost without exception (Sustainable Okotoks, Public Participation Process, 2005). For Ms Berry, the checks and constraints built into the sustainability plan are sufficiently well-entrenched in the local community and its operating culture that the erosion of community sentiment is unthinkable and the appearance of a winning no-bars-to-growth political coalition is only slightly less unlikely.

In a subsequent telephone interview, former municipal manager Will Pearce (2005) reinforced Ms Berry's optimism. Concerning the dangers of development momentum during the "build-out" period overflowing the accepted limits to population growth, Pearce offered a number of safeguards that militated against such an eventuality. First, he pointed to the extremely supportive statutory framework of the Province of Alberta, which lacks provincial mechanisms to trump local efforts at devising autonomous or self-directed planning. Alberta's provincial offices, he noted, stand in sharp contrast with the Ontario Municipal Board, which oftentimes quashes local initiatives and self-directive policies that run counter to regional or provincial development policies and programs. He did acknowledge that a progressive-minded group within the council or a citizen action group who put community needs over self-interest would perpetuate or enhance those efforts he had initiated.

Beyond these facilitating provincial and local conditions, he highlighted his and his planning team's cognizance of how front-end controls on longer-term development pressures and risks might overwhelm initial community support for the local sustainability plan. For one thing, while working to mobilize broad-ranging community buy-in into the program, Sustainable Okotoks planners aggressively pursued business support. Among other constituencies that they courted, planning officials sought to convince the town's real estate developers and contractors that moving from R01 housing to mixed housing would (and eventually did) increase profits from sales of diversified houses to a wider range of potential buyers. He also noted that periodic surveys were generally performed a year before municipal elections and the results (always supportive of the vision) were released and publicized a couple months before election-day voting to influence the outcome of elections. Because council candidates were made aware of community sentiments supporting the program, the lure of rampant development campaigns was minimized and as a result, progrowth candidates almost invariably fared poorly in local contests.

In addition, various planning actions were implemented to thwart leapfrogging the growth boundary through development outside the municipality's borders. One major step was to downsize the publicly supported infrastructure of development within the city's boundary so that public infrastructure (utility lines, water and sewer pipes, roads, etc.) did not, in effect, subsidize development outside the growth envelope. For example, once the Sustainable Okotoks plan was adopted, the size of water and sewer pipes was reduced as the lines approached the town limits. Developers could no longer tap into oversize pipes, already laid by the municipality, thus imposing an added financial burden on any land speculator by forcing the developer or contractor to pay for oversized piping as a cost of new development outside of the municipal boundary. Another example (already alluded to) was the negotiation of a 20-year (not a 5-year) intermunicipal development plan with the town of Foothills – one that fostered clustered rural development around Okotoks. This practice also obviated Okotoks' need to provide water and sewer to rural development projects in Foothills (by checking the expansion of commercial and residential developments that could not meet established Foothills' rurally zoned regulations) (Pearce, 2005).

Whether these views amount to complacency, hubris or reasoned self-confidence borne of solid evidence is uncertain. The record of urban sustainability programs in North America suggests that while Canadian towns and cities like Hamilton-Wentworth, Montreal, Toronto and Bamberton may not have found the winning formula for fully embedding strong sustainability into their urban planning and policymaking processes, they have succeeded in dreaming of it. They have also implemented a variety of sustainability projects that are better than what has been accomplished by their American counterparts. As Hamilton-Wentworth's Vision 2020 program has demonstrated, Canadians have benefited from a strong and intelligent civil service and from an apparent trust in their civic traditions and political culture to act patiently and work tirelessly – for the long haul – to realize future goals (Yanarella, 1999). Simultaneously, that record is sufficiently checkered and progress is likewise subject to shifting political fortunes so that complacency and self-satisfaction are not warranted. Recognition that policy reversals have occurred elsewhere should prompt public officials to work hard *now* to build up reserves of public goodwill and support for Sustainable Okotoks and find ways to root its animating purposes in the larger community and its institutions of civil society, lest misfortune happen. Many a well-intentioned public program has been overturned by "outrageous fortune" and a tenacious, well-financed and single-minded coalition of individuals whose private interests and shortsighted vistas were substituted for the common good.

The self-assured public officers of Okotoks would do well to begin running scared and looking over their shoulders for signs of eroding political support and rising oppositional elements. These are likely to be generated by changing demographics and conflicting interests and cultural lifestyles arising from population trends noted above, which may dilute or substitute for that grand vision animating Sustainable Okotoks at its creation in 1998. For not only is the adage that "misfortune happens" an enduring one; but no less longlived is the maxim that "the work of politics is never done." And with Will Pearce's departure to Cranbrook, British Columbia, Okotoks must depend upon the strength of his legacy and the continuing commitment of its citizenry and business community to its lessons.

Rumble from the North: The Unraveling of Sustainable Okotoks?

For ten years, Sustainable Okotoks benefited from with overwhelming town leadership and citizen support, and experienced a virtually unbroken number of successes in meeting or exceeding its environmental targets. The mayor and town council shaped its downtown development and infrastructure needs around the underlying vision. The citizenry bought into its water conservation program, recycling requirements, land-use policies and greenhouse gas emissions reduction targets. Outside organizations, professional associations

Table 9.1. Umbrella/core sustainability planning accomplishments

Umbrella plans and agreements	Okotoks	Town of Okotoks Strategic Plan (September 1995) This plan contains: • The Town of Okotoks 2010 Community Sustainability Vision • The Town of Okotoks Corporate Mission
	Regional	The Town of Okotoks/MD of Foothills Intermunicipal Development Plan (September 1998) MDP reviews: • 1998–2003 • 2004–2008 • 2007–2008
	Okotoks	Town of Okotoks/MD of Foothills Intermunicipal Development Plan (September 1998)
	Regional	Town of Okotoks' four foundations of sustainability: 1. Environmental stewardship 2. Economic opportunity 3. Social conscience 4. Fiscal responsibility
	Okotoks	South Saskatchewan River Basin Water Management Plan: • Phase 1 – Water Allocation Transfers (June 2002) • Phase 2 – Final Plan (August 2006)
	Regional	Calgary Metropolitan Plan (June 2009)
	Okotoks	Sustainable Okotoks – Where To From Here? (SOWTFH) (2009 – current). Community consultation via publications, resulting in the 2009 community survey

Source: Town of Okotoks Planning Department

and political bodies conferred sustainability awards and other accolades upon Sustainable Okotoks for its notable achievements. Beginning in 2008, the integrity and longevity of the 1998 Sustainable Okotoks' Legacy Plan and growth management policies were shaken by three significant forces that threatened to overturn its guiding framework.

The first challenge came from the Province of Alberta's decision to mandate regional planning through the formulation of a provincial Land Use Framework, targeted for adoption in April 2009. Since Okotoks is situated within one of the seven planning regions comprising the province (i.e., the South Saskatchewan Planning Region), its land use and other connected policies fell into this regional planning framework and, as a result, would be affected by Alberta's requirement to craft a Calgary Regional Land Use Plan (CRLUP) along with the other CRLUP member communities. Complicating this regional planning initiative was the decision-making structure that would give Calgary the dominant voice in shaping the details of the CRLUP by creating a "supermajority," where 12 of 17 CRLUP community members and 50 percent of the region's population would constitute a "supermajority" for making binding decisions where consensus among the membership was lacking. Because Calgary's population constitutes 80 percent of the region's population, its opposition to any other member actions could effectively veto any proposed action (for an overview of the Alberta Land Use Act and CRLUP, see Quail, 2009).

The galloping growth in population and residential development of communities surrounding Okotoks, especially the Municipal District (MD) of Foothills, was a second complicating factor posing problems for Sustainable Okotoks' objectives and self-imposed constraints (Markusoff, 2009). Despite Okotoks' explosive growth over the preceding ten years, its interrelated population and water usage constraints brought home to diverse community constituencies the realities of a decision forged when Okotoks' population was scarcely above 11,000 inhabitants. If Okotoks held fast to its sustainability yardstick and did not grow beyond the 25,000–30,000 population limits, it would likely be faced with rising growth in its surrounding communities and pressures to hook into Okotoks' services or accommodate the infrastructure needs of those nearby growth areas.

Odd though it may seem, the third destabilizing force was the approach of the population cap of 25,000–30,000 inhabitants, stemming from the town's water license limit of 604 million gallons per year. Having used approximately 580 million gallons of water in 2009, Okotoks' 21,690 residents (2009 est.) would likely meet its water limit by 2011 and with a population well below 30,000 (Patterson, 2010a, b and c; and Barlow, 2010). As various stakeholders became increasingly aware of the ramifications of this barrier, some stakeholders became uneasy about what it would mean for the community and themselves. The Okotoks' Chamber of Commerce began to express mounting alarm at the antigrowth consequences for businesses wishing to move to Okotoks (Markusoff, 2009). The long-time mayor of Okotoks Bill McAlpine and several town council members also expressed unease with this happenstance, especially in light of mounting pressures from the regional planning body and the province to induce Okotoks to align with regional and provincial policies related to anticipated population increases and water demands (ibid.).

The combination of these three elements triggered a widespread debate within Okotoks' community and prompted the mayor and town council to launch a public information and consultation campaign to educate Okotokans about the available alternatives to meet these new forces and to consider the short- and long-term implications of each. In that public education drive, a flurry of information sessions were scheduled that began with news releases published in the weekly community newspaper, *Western Wheel* (collected as links: Town of Okotoks, 2009a). These information bulletins carefully outlined the strongest case for each of three options around which different stakeholders and camps had formed.

One option, called the "red light" or status quo scenario, was offered as the best case for retaining the original 1998 legacy plan. A second alternative, dubbed the "green light" or CRLUP scenario, made the strongest case for Okotoks to buy into the evolving regional plan, which would shelve the limits of the Sustainable Okotoks program, grow the city's population to approximately 60,000 inhabitants, accommodate expected aggregate regional growth of 1.6 million and build a regional water pipeline from Calgary to satisfy the needs of the augmented local population. A third option, known variously as the "yellow light" or tweaked scenario, was favored by the mayor and his supporters on the town council, and would necessitate town growth to around 45,000 inhabitants, require increased water capacity and lead to annexation of surrounding land (summaries of these three positions may be found in Quail, 2009).

In addition, other information sessions involving representatives from the Calgary Regional Partnership (CRP) were conducted and at least two public comment sessions were held to encourage widespread commentary by town supporters *and* opponents of the three alternatives.

During this information and consultative process, two citizen action groups surfaced: Citizens for a Sustainable Okotoks and NoCalgaryVeto. These small, fledgling groups mounted petitions, developed webpages, wrote letters to the local newspaper, posted to government websites and provided testimony at public sessions. Both sought to defend Okotoks' local decision-making authority and autonomy, and each strongly opposed the erosion or upending

Table 9.2. Red, green and yellow scenarios

MDP growth management	The three scenarios compared to 1998 MDP		
	Status quo (Legacy Plan) 2019 – 30,000	Tweaked (modified growth boundary) 2039 – 45,000	CRLUP (full growth) 2076 – 60,000
Clear targets negotiated in advance	Town achieved locally	Town and CRP achieved in regional role	Town and CRP achieved in regional role
Gradual, managed evolution of land use	Does not deal with population growth pressures on town's boundaries. Achieves Legacy Plan	Phasing possible based upon water capacity. Can be achieved	Phasing possible based upon water capacity. Can be achieved
15–20 year build-out	Legacy Plan will achieve build-out but no long-term projected population (CRLUP)	Accommodates half of the long-term projected population (CRLUP), an additional 15,000 people	Accommodates full long-term projected population of an additional 30,000 people
No significant annexation	117 ha (290 acres)	300 ha (730 acres)	675 ha (1667 acres)
Density targets for new development	4.0	5.0 to 8.0	5.0 to 8.0
Sustainable neighborhood design	Least transit-oriented design	Moderate transit-oriented design	Most transit-oriented design
Comprehensive community plans	Meets Legacy Plan and some CRLUP objectives	Incorporates majority CRLUP objectives	Meets the CRLUP objectives

Source: Quail, 2009

of the 1998 legacy plan (for information and blog entries, see the Citizens for a Sustainable Okotoks and NoCalgaryVeto websites). Though the membership of these organizations was small in number, their message clearly resonated throughout the community. This was evidenced by the April release of the 2009 annual community survey results just prior to this issue being put on the 25 May 2009 town council meeting. Those results demonstrated that: 55.9 percent of Okotoks households supported maintenance of the status quo – i.e., retaining the legacy boundaries and population and water limits; 25.2 percent endorsed the tweaking option, supporting moderate boundary growth and subregional or local water solutions to address increased water needs; and 15.9 percent preferred the CRLUP plan, endorsing growth to approximately 60,000 inhabitants and allowing a much larger growth boundary (including a regional pipeline solution to address heightened water consumption flowing from the augmented population) (Town of Okotoks, 2009c).

This controversy was temporarily settled by the Okotoks' town council at its 25 May 2009 meeting when it voted for a two-year delay before rendering a final decision on whether or not to overturn the population cap and move toward a geographically expanded community with more residents (Markusoff, 2009; NoCalgaryVeto, 2009; Town of Okotoks, 2009b). In addition to the vocal protests of the two citizens' groups, the town's elected officials acknowledged the political sway of continuing majority support by residents for retaining basic components of the original 1998 legacy plan and minority support for the other options expressed in the 2009 community survey (Markusoff, 2009). On the other hand, the council passed a resolution to work in partnership with the MD of Foothills to achieve a subregional growth management strategy and agreed to move forward with the CRP and proposed amendments to the Metropolitan Land Plan. These three actions suggested that the town council's "compromise" among competing interests amounted to a Solomonic act that postponed immediate shelving of the framework of Sustainable Okotoks' key principles, but opened the way for the CRP governing process to nullify Okotoks' standing limits on population growth and the water cap – and effectively to impose its proposed regional plan upon Okotoks and other CRP member communities in the near future (NoCalgaryVeto, 2009).

Some Policy Recommendations

Having embarked upon a pathway to local sustainability between 1998 and 2008, the Town of Okotoks has come to a juncture in the first phase of its sustainability journey; it now needs to reevaluate seriously its overall vision of a sustainable future and the various elements it has put into place for realizing that vision. Its local resources of farseeing public servants, effective administration and commitment to the long term are readily apparent, although the council actions on 25 May 2009 have put in question whether these local assets, which have emerged from its distinctive history, traditions and attachments to place, can remain intact and prove effective in repelling outside pressures for unsustainable growth.

The Sheep River has been something of a unifying symbol to the underlying vision of a sustainable future and the carrying capacity of its waters has been used for over a decade to impose a cap on future population growth. As Okotokans turn their eyes and thoughts forward to the evolving landscape of their community and the new dynamics of the region and the province, it seems clear that they will have reached the second phase of their journey to sustainability. There, they will have to draw upon native resources (again) to reshape and refine that vision to take account of the implications of the policies and decision making which informed the first phase of implementing Okotoks' strategic vision.

Though the elegance of their solutions to the cultural, transportation and other challenges ahead will no doubt be predicated upon their unique appreciation of their sense of place, the citizens of Okotoks have many valuable insights and important lessons – both positive and cautionary – to share with other small towns and communities in Canada, the United States and elsewhere. As they forge new tools and successful solutions to both old and inchoate problems that stymie lofty goals, the authors offer the following set of policy recommendations.

Encourage the emergence of institutions of civil society focusing on all dimensions of sustainability. Between the economy and formal government lies the welter of civic organizations shaping and enriching the everyday lives of inhabitants as members of a community. A survey of Okotoks' social landscape would likely show a dense and overlapping network of social and civic groups inhabiting its civil society. Conspicuously absent, at least until recently, has been the formation of such independent groups and organizations structured around or contributing to the evolution and evaluation of the town's sustainability programs. It may be tempting in a country like Canada where the tradition of civil service and local management is strong – and where its administrators and planners are, as a rule, better educated, more cultured and politically savvy than their American counterparts – to depend upon these instruments of good government and efficient administration. If the largely top-down orientation of Sustainable Okotoks is to be softened and balanced by strong community-based civic organizations that can root its objectives and promises in community life, such citizen agencies of democratic voice and constructive political change must be encouraged, cultivated and even tolerated by local officials and program administrators. The emergence of Citizens for a Sustainable Okotoks and the NoCalgaryVeto groups in the face of regional and provincial assault on local autonomy and home rule are surely positive signs of the flowering of institutions of civil society. But these small citizen groups will have to be fortified by greater numbers and a wider regional and perhaps even provincial vision of sustainability. Otherwise, the legacy of Sustainable Okotoks may be consigned to history's dustbin by the regional and provincial growth imperatives bearing down on this small southern Alberta town.

Lay the groundwork now for a coherent and comprehensive transportation policy that anticipates the impact of area residential and commercial growth. The municipal development plan inaugurating Okotoks' sustainable program restricted future residential growth to accommodate the prescribed upper limit of population increase to 50 percent north of the downtown and 50 percent to its south. In so doing, it created two poles of evolving residential growth. Whatever the scale of the growth management policies that emerge out of the latest contest among local, regional and provincial forces, if future transportation policy is not to become hostage to the private automobile and to the seeming imperatives that drive road building and transportation decisions, local elected official and planners must think through the wisdom and value of future public transit. So far, the only evidence of transportation planning has come from CRP deliberations regarding intermunicipal express bus and rail services (Guttormson, 2008).

A central issue for discussion and planning is how plans for future public transportation (a bus line or faux trolley service) will impact road construction through the edge of residential development in the northeastern pole of suburban development (along 32nd Street) and the southern pole on Southridge Drive across from the shopping plaza anchored by Walmart.

How downtown revitalization figures in such transportation and road planning is also another key question for investigation and debate. While not populated by empty street-level shops, it is our impression that the survival of the downtown corridor as a magnet for local citizens remains precarious. Future road construction and public transit planning are two areas where citizens, elected officials and representatives and town administrators need to open up a community conversation, lest the crush of more automobiles and the consequences of lifestyle preferences of a changing demographic profile force policy decisions contrary to an expanded and more coherent sustainability design. As Pearce (2005) acknowledged, the Municipal Development Plan was based on the general guideline that the sustainable community be designed so most inhabitants can walk to meet their daily needs and have easy access to a pathway system that facilitates that goal. If, as he further noted, projections and cost estimates indicated that the "tipping point" for a reasonably subsidized public transit system for Okotoks would come when the population reaches

20,000–25,000 inhabitants, it behooves the mayor, council and planning offices to lay the groundwork for that fast-approaching day.

Foster the continued development of a culture of sustainability. An important bulwark against flagging community enthusiasm and general backsliding in political support for Sustainable Okotoks' long-term aspirations is a robust and vibrant culture of sustainability, which takes many expressions. The pioneers of the local sustainability program intuitively sensed its importance by supporting the conversion of the old train station into an arts and crafts center and the translation of historic preservation sensibilities into a government-supported heritage and archives facility. These fledgling operations show great potential to expand their agendas and to dovetail with aspects of the town's sustainability initiatives, as the recent arts and crafts exhibition for community artists by the Cultural Station and the 2005 flood exhibit at the Heritage House foreshadow. Cultural projects along these lines will need to be augmented and reinforced by community education programs, which target particular age groups and are already evidenced by Okotoks' garbage collection and recycling programs and their periodic riverfront community cleanup activities. Okotoks might consider some of the public education programs that were instituted by the Hamilton-Wentworth Vision 2020 program, such as its annual Sustainable City Day celebration and its Citizen Summits. These techniques would help to maintain a continued dialogue between citizens and public servants who are on the course of local sustainability. In this respect, its public education activities must be directed to children and youth and its other programs periodically revitalized to rejuvenate interest, enthusiasm and community understanding of the animating purposes of the original program. Still, while these and other innovative and successful ideas can enhance local efforts, a culture of sustainability will best be found in the commonsense wisdom, indigenous resources and practical insights borne of its citizens' love and appreciation of a place called Okotoks.

Work to transform the downtown corridor from a place where there is no "there" there to a center crowned by a public square. A troublesome shortcoming of unfolding plans to revitalize the downtown of Okotoks is the absence of a town center or a public square from which the rest of town radiates. True, the town's municipal development plan (Town of Okotoks, 1998: 77ff) outlined an agenda for rejuvenating the economic and public face of the old town center. A commissioned study by Calgary-based HABICO Planning + Architecture Ltd generated public space design guidelines that assisted in rethinking downtown design issues and fashioning possibilities for public space within the historic core (HABICO Planning + Architecture Ltd, n.d.). Still, as the 1998 MDP document admitted, "downtown revitalization efforts...have a short and frustrating history" (Town of Okotoks, MDP, 1998: 77). Moreover, the linear character of Elizabeth and North Railway streets presently militates against the idea of a centrally located and pedestrian-friendly public square where the town's inhabitants are daily drawn or regularly assemble for community festivities and celebrations.

The initial impression gained by a visitor who tours the town by automobile is that the historic tendency – and the current trend under the present growth regime – has been an entropic plan resulting in scattered community centers with commercial and residential development further and further away from the downtown anchor. On the other hand, the old town area bounded by the new library, the town hall, the Heritage House and cultural station suggest incipient possibilities for redevelopment that could produce a recentralized focus to the town. We would argue that for a sustainable future, Okotoks needs a public square to realize its plans and ambitions and must stitch together the residential development in the northeast along Highway 32 with the commercial and residential growth along Southridge Drive.

As suggested above, a carefully planned and town subsidized public transit system which takes passengers through the historic town center could play a critical role in bringing people to (and through) the downtown in a manner that would boost the downtown's business community and enhance the entire town's civic life. New and innovative visioning and planning are badly needed. Revitalization schemes bent merely on economically improving downtown businesses are far too shortsighted. Confined public spaces without a civic purpose are simply not enough.

Work to integrate the diverse parts of Sustainable Okotoks into a whole policy cloth. A visitor to Okotoks cannot help but be impressed by the many policy initiatives deployed by the town's government and administration since the formulation of Okotoks' sustainability strategic vision. In spite of its central thrust (building up only to its water supply's maximum carrying capacity) and its many components, Sustainable Okotoks strikes one as a community program whose individual threads have not yet been woven together into a single tapestry. A recycling ethic generated by local regulations seems to have become rooted into local culture. Solar energy has been showcased in a number of individual projects using advanced technology and best practices. Wastewater treatment has been upgraded and informed by a conservation and energy efficiency sensibility. However, a sense of wholeness and coherence of the parts seems to be lacking. Moreover, one missing piece of the sustainability picture puzzle is the roads and transportation component. Part of the task of bringing coherence to the town's sustainability vision is to think of each of these elements in relationship to the other. For example, how do solar energy and conservation projects contribute to sustainable water policy? How can instances of cultural sustainability contribute to strengthening an ethic of resource conservation and recycling? These issues promote the most searching policy inquiries and the widest possible community dialogue. Not only does the future of Sustainable Okotoks depend upon answers to these queries, but in a fundamental sense the shape of Okotoks and its small-town character hang in the balance.

Utilize the technique of multiple or alternative scenario building as a way to overcome win-lose outcomes in the politics of local-regional sustainability. The "compromise" actions of the 2008–09 public imbroglio over the three alternatives that occurred in the 25 May 2009 town council meeting bears all of the earmarks of a classic interest-group liberal resolution, where something was given to all parties and the most difficult choices were deferred. This political context (and perhaps others in the future) might have been better settled by using a more collaborative process where multiple scenarios were pursued as a Sustainable City Game (see Chapter 6). What might have resulted? We submit that more creative and even unexpected outcomes, which merged elements of the best recommendations flowing from playing out each scenario (on multilevels) and the introduction of acceptable elements from each would have synthesized into a broadly agreeable resolution, unanticipated by hardcore interests and attitudes of involved stakeholders. It should be borne in mind that the originating vision of the Land Use Framework was that:

> Albertans work together to respect and care for the land as the foundation of our economic, environmental and social well-being and its putative outcomes are intended to be: a healthy economy supported by Albertans' land and natural resources, healthy ecosystems and environment, and people-friendly communities with ample recreational and cultural opportunities." (Province of Alberta, 2008)

This vision and those valued outcomes are laudable ones and, interpreted in twenty-first century terms, reflect global necessities that can be convergent with strong sustainability. As one of our five operational principles of sustainability suggests, "all lower imbalances are to be negotiated outwards," which is to say that sometimes local imbalances can be restabilized – through creative and more encompassing democratic participation and policy actions – at a larger (perhaps) regional level. Thus, the specter of a sustainable Okotoks – which is surrounded by a contiguous community of heightened population growth, with commercial and residential development and wider regional pressures from population in-migration throughout the rest of the Calgary region – might be the only way to institutionalize strong sustainability over the long run.

Conclusion

One of the ironies of Okotoks' early twentieth-century economic history is that for nearly four decades it was the gateway to the nearby Turner Valley oilfields (see below). From 1913 to the early 1950s, Okotoks' economy was heavily invested in supplying and transporting a wide range of oilrig equipment and secondary products to the fields. As a result, its economic fortunes were importantly tied to the vicissitudes of the regional – and international – oil industry. Today Okotoks is forging a way for its citizens (and more recently for Calgary, the metropole to its north) to move beyond diminishing fossil fuels and profligate energy and other natural resource consumption into the Age of Sustainability.

Much of the original enthusiasm that set Okotoks on a path toward sustainability remains embedded in community sentiment and institutional practice, even as outside forces threaten to lay siege to its hard-won accomplishments. But questions and doubts have begun to surface within the local body politic. In the coming years, more will be called upon its concerned citizenry and its dedicated town leaders and administrators – if the powerful forces operating in their midst and gaps in their sustainability blueprint are not allowed to divert this bold experiment from achieving its highest aspirations. In juxtaposing the telos of Okotoks' sustainable vision against the pressing realities and possible stumbling blocks that are now becoming apparent, this analysis and critique have been buoyed by hope in this enterprise and confidence in the ability of Okotokans to become engaged and to overcome these challenges looming on the horizon.

Chapter 10

VIENNA'S WESTBAHNHOF SUSTAINABLE URBAN IMPLANTATION – THE CITY-AS-A-HILL

Introduction

The new conditions of the twenty-first century beckon us with a future offering both enormous promise and unparalleled risk. Like the age of Dickens' two cities, the twenty-first century presents opposing faces. It will be either the century of sustainability or the century of ecological collapse. It will be the century where the continuation of the unsustainable economic practices of today precipitates irreversible catastrophes. Or it will be the century where small local successes in implementing sustainable practices and processes proliferate and transform the global economy into a balance-seeking relationship with our natural ecosystem. It may be the century where the analytical, reductionist methods of science and industry, which are the sources of both our progress and our increasingly unsustainable way of life, will continue as the central economic paradigm. Or it will be the century where a new integrative economic paradigm emerges which promises to reconcile humankind with the natural environment, whose health is the precondition for all human activity.

We are confronted with ecological and social choices of continuing our descent into the realm of unsustainability (denying all the while that it is the consumerist-materialist path that is the problem), or shifting to a new emerging paradigm. Unfortunately, in order for massive popular support to emerge to accomplish such a paradigm shift, there must first be a widespread belief that a larger, attractive, sustainable, new societal project exists and can succeed. Apart from invoking the all-too-familiar litany of "low-hanging fruit," which promises only to slow the rate of descent, few recognize that a substantial alternative exists to the prevailing path. Therefore, the primary challenge facing our generation is to develop a real and viable alternative to define the sustainable city, not merely on a theoretical basis, but in a real place.

Urban sustainability is an idea whose time has come (see Chapter 2). While the required change may seem enormous, it may actually take relatively little to precipitate such a paradigm shift. It took only one steam engine, one light bulb and one photocopying machine to trigger social and technological revolutions. So too, it may take only one city operating within the limits of its environmental resources to prove to all others that not only is sustainability possible, but that it is the only probable way for cities and their economies to be designed and managed to ensure their long-term survival. Fortunately, there is an operational theory of the sustainable city and it has been embedded in the European Charter of Cities and Towns Towards Sustainability (European Commission, 1994). Many towns and cities are attempting to implement the European Charter in their programs and urban projects.

The focus of this chapter is Vienna's Westbahnhof project, an ambitious undertaking commissioned by the City of Vienna and the National Bank of Austria over a five-year period (from 1994 to 1999). This project was jointly developed by Oikodrom – known as the Vienna Institute of Urban Sustainability – and the Center for Sustainable Cities at the University of Kentucky in Lexington. Because the underlying framework of this project departs in significant respects from conventional approaches to sustainability, we begin this chapter with a description of its theoretical underpinnings and design elements, as well as the historical model and precursor to its "city-as-a-hill" design. Then, through text and illustrations, we explore the fundamental efforts to integrate urban architectural design with strong sustainability principles into a program for implementing a sustainable city implantation (SCI) upon and over the present Westbahnhof site. The purpose is to demonstrate not only that sustainability in a place is imperative, but also that it is possible and realizable. The chapter concludes with some important lessons derived from the program's ambitious vision and ultimately its political fate. Had the project been constructed, it would have become the first project in Europe to fully implement the principles of the Aalborg Charter and in so doing become Europe's first modern sustainable city.

The Sustainable City of the Past

The sustainable city is not a new phenomenon. Historic towns and cities around the world did not have a choice if they were to survive. If they succeeded in existing for any length of time, it was only because they were able to develop and maintain a continually rebalancing relationship among their internal social and economic activities and with their wider natural and agricultural landscape (Levine, 1987). Ironically, it is a tribute to our genius that we have created an artificial economic system and a larger "second nature" that have operated on an unsustainable basis for many generations.

We have been able to do this because of the brilliant ways in which we have contrived to export the problems our expansive way of life has created either to poorer, less defensible regions, or to future generations. However, the fact that the consequences of those problems cannot be indefinitely forestalled and that the cascading effects of that unsustainable future are rapidly mounting is becoming increasingly clear.

The sustainable patterns of past cultures are very similar to the ways in which a natural ecosystem operates. Like an historic town, a natural ecosystem is a strong local economy working as an interconnected network typified by regenerative cycles of energy and material flows. In nature, processes are cyclical and regenerative; there is no waste. Over time, the diversity, resiliency and thus stability of the ecosystem increases. As Redclift (1987: 18) shows, human society typically seeks to stall such ecosystems in early stages of ecological succession, leading to high production without the accumulation of some equivalent of organic matter and biomatter to promote stabilization. High production within these human-built ecosystems comes at a high price. As discussed earlier in this work, the result is a state of arrested development in social, environmental and even economic systems, which are marginalized when they become secondary to the immature commitment to low resiliency caused by the seeming privilege of limitless quantitative growth.

The good news is that the sustainability movement is growing and it is a worldwide phenomenon. However, even with this broad movement, there is no consensus as to what sustainability effectively or operationally means. The Earth Summit in Rio in 1992, the Climate Change Conference in Kyoto in 1997 and the Copenhagen Conference in 2009 are just a few of many international activities confronting our growing environmental problems. The bad news is that the principal vehicles employed in this movement are the largely analytic, reductionistic methods that have arguably created the very problems that the movement is struggling to resolve. This overwhelmingly quantitative approach insists that we must commit ourselves to do more with less and less. While this approach offers success in its initial steps (recycling aluminum, changing to fluorescent fixtures, more tightly insulated homes and so on), subsequent steps become increasingly more difficult, more expensive, less popular and less effective – until a point of diminishing returns is reached long before meaningful reductions, much less sustainable balances, have been achieved (see Chapter 4). Equally troubling is that this pervasive quantitative approach is also a top-down approach starting at the global level (e.g., the Rio Summit) and working its way down to regional, national and local programs of restriction and regulation. Such an approach is fraught with inevitable controversies and conflicts (i.e., who is to make which sacrifices and on what basis?) that are likely to breed dissension and division and defuse any preliminary momentum. In any case, as the Rio Plus Five conference indicated, the record has not been promising because the nations of the north have been unwilling to be strong role models to their neighbors to the south. An alternative approach exists that, while less visible in the United States, has achieved greater momentum in Europe. A major embodiment of this approach is seen in the European Charter of Cities and Towns Towards Sustainability (the Aalborg Charter) which, in contrast to the reductionist approach, is holistic, process oriented and place-centered (see Appendix A). It takes the city-region as the appropriate scale for analysis and action. As noted in earlier chapters, this approach removes the problem from a strictly scientific, analytical framework – it mobilizes a whole realm of social science and civil society issues and demands a broader process approach. Such an approach can only be fully addressed in the realm of transdisciplinary processes and architecture and urban design. By its very nature, this approach promises to extend both the nature and influence of those realms. As principal architects of the Aalborg Charter, we are pleased that hundreds of Europe's most progressive cities have ratified this document, evidencing the commitment of those municipalities to create and implement action plans based on the Aalborg Charter.

At least since the late 1980s, the sustainability movement has been organized around a minimalist, consensus definition advanced in the United Nations through the Brundtland Commission's report, *Our Common Future* (UN, WCED, 1987). Instead of this rather nebulous collection of good wishes, we propose an alternative course: a robust theory and process for generating and operating a sustainable city-region. It is based on the Aalborg Charter, which in turn derives from the "Five Operating Principles for the Sustainable City" and our alternative definition of sustainability. (See Chapters 2 and 5 for earlier elaboration of the components of this alternative strategy.)

The Westbahnhof Project – The Sustainable City of the Future

The city of Vienna, Austria, remains beset with a major urban problem: a 1.4 km long rail yard near the heart of the city that slices the 15th district in two and creates substandard living conditions along its length. Instead of conventional overbuilding (*uberbaung*) of the rail yard (as may be found in many other European cities), then vice-mayor and director of urban development Hannes Svoboda commissioned a study for a sustainable urban implantation (SUI) on the site according to the principles using the new urban city-as-a-hill concept. Developed conceptually through numerous architectural design studio projects and field studies, the SUI was inspired by the historic medieval European hilltown. (See figures 10.1, 10.2 and 10.3.) This city-as-a-hill

Figure 10.1. Westbahnhof railway overview

Source: University of Kentucky Center for Sustainable Cities and Oikodrom

Figure 10.2. Photomontage of proposed Westbahnhof sustainable urban implantation

Source: University of Kentucky Center for Sustainable Cities

prototype, rendered through the Sustainability Engine, presents a new holistic, people-centered urban vision. The Sustainable Area Budget (SAB) method generates a great deal of its power by avoiding the fruitless exercise of trying to establish an arbitrary standard for reducing the degree of unsustainability that prevails. In such an exercise, the degree of unsustainability is reduced by an assumption about what is presently economically and politically feasible – not by what is *actually* sustainable. Unfortunately, this is an exercise that is widely performed in the name of sustainability. In contrast, sustainability is nonnegotiable in the SAB process. This precondition sweeps away numerous irrelevant issues and greatly simplifies the entire process.

Theory and practice – Form and counterform

Having established a coherent, consistent and complete theory of the sustainable city-region, the question remained how to proceed from theory to practice. In fact, the evolution of the theory has paralleled the coevolution of a new urban form. In seeking an appropriate form and structure for the sustainable city of the future, the aim is to synthesize a new model where problems of the modern city never appear, rather than identifying the means for solving problems of existing cities. At the same time, when the process definition of sustainability is strongly adhered to, it can be a major producer

Figure 10.3. Second iteration of Westbahnhof SUI

Source: University of Kentucky Center for Sustainable Cities, Richard S. Levine and Taqi Radmard (architects)

of urban form and organization. In the iterative trial-and-error process known to all designers, numerous different concepts can be proposed and studied. Unproductive directions can be abandoned and promising models saved. By circling around the problem over and over again, a locus of mutually supportive relationships slowly emerges. It is not a question of seeking a "perfect form," but of determining and developing a family of forms with the flexibility to accommodate a variety of local preferences.

Thus, structures are chosen at a variety of scales which have mutually supportive tendencies and can be associated with the needs and possibilities of sustainable cities. From our definition of the sustainable city-region, one can infer a number of strong tendencies (Figure 10.4). The definition supports a dense, compact city with a dynamic balance between community and privacy; it supports a community rich in form, public space and individual and collective opportunities; it supports a city with a strong sense of itself as a place, with a clear and well-defined form and a common destiny. The definition supports a human-scaled environment, not one that is overscaled or sized to accommodate vehicles, industry and faceless institutions. Yet it

Figure 10.4. S-PATH model rendering, Korean Public Administrative Town Competition

Source: University of Kentucky Center for Sustainable Cities Design Studio, Richard S. Levine (architect) and James Black (renderer)

also supports a city able to provide appropriate space for various larger-scaled industries and the infrastructure necessary to accommodate the metabolic and economic processes of a modern city.

The City-as-a-Hill – A New Urban Model

In seeking to create an urban structure with the above characteristics, our proposal for an innovative type of city district combines some of the most compelling aspects of the medieval European hilltown with the best of modern processes and technology. Instead of the medieval city-on-a-hill, we propose a city-as-a-hill (Figure 10.5): a construction whose outer surface resembles, in scale and texture, the human-scaled medieval towns. Using advanced computer modeling software, which allows for the possibility of generating many varieties or models of SUIs in an interactive and participatory manner, this distinctive urban configuration creates many opportunities that would be impossible in the modern-day unsustainable city (Levine and Radmard, 1990; Levine, Yanarella and Radmard,1991). In our city-as-a-hill model, the outer surface of the city contains all of the dwellings and neighborhoods, the smaller-scaled commercial and institutional activities, plus the network of public buildings and public spaces – that is the streets, walkways, stairways and center squares, which give historic medieval towns their life-affirming pedestrian character (Figures 10.6, 10.7 and 10.8).

Inside the city-as-a-hill, daylit by courtyards and light wells, a series of three-story concourses and gallerias stand, alongside large-scale commercial, institutional and industrial spaces, as well as the infrastructure necessary to support them. Over the years, the structure and complexity of these models

Figure 10.5. City-as-a-hill cross-section

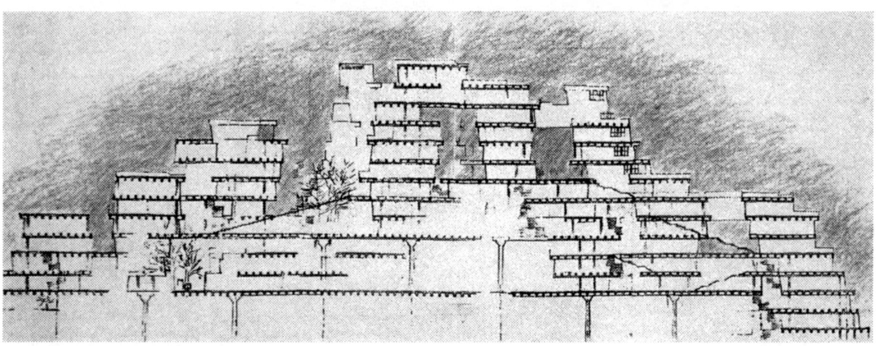

Source: University of Kentucky Center for Sustainable Cities Design Studio, Richard S. Levine (architect) and Taqi Radmard (architect and renderer)

Figure 10.6. City-as-a-hill model in practice (Whitesburg, KY)

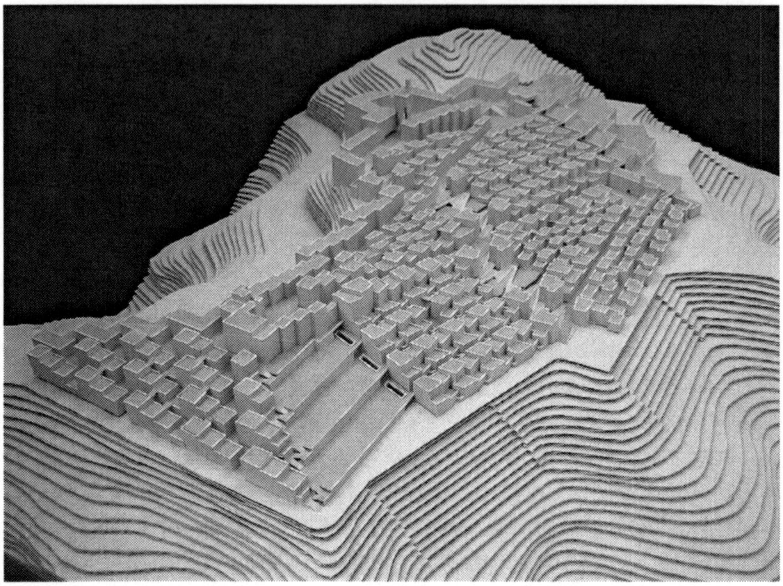

Source: University of Kentucky Center for Sustainable Cities Design Studio, Michael Hughes and Casey R. Mather (designers)

Figure 10.7. Urban landscape of city-as-a-hill urban model

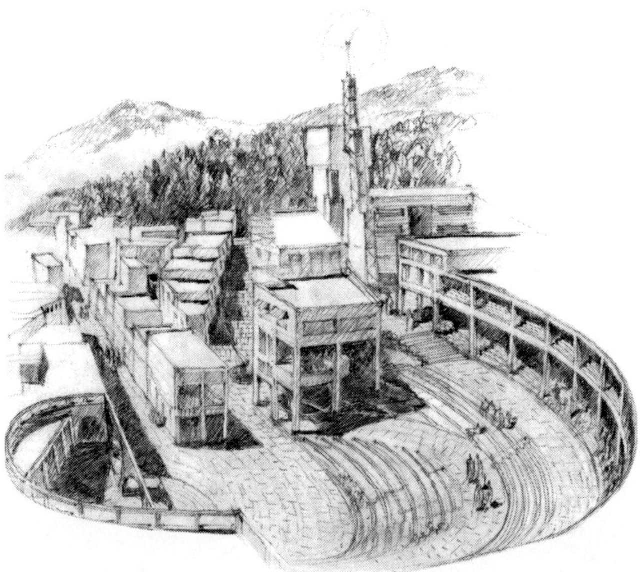

Source: University of Kentucky Center for Sustainable Cities Design Studio, Taqi Radmard (renderer)

Figure 10.8. Sloped street intersection of city-as-a-hill model

Source: University of Kentucky Center for Sustainable Cities Design Studio, Taqi Radmard (renderer)

Figure 10.9. Coupled Pan Space Frame

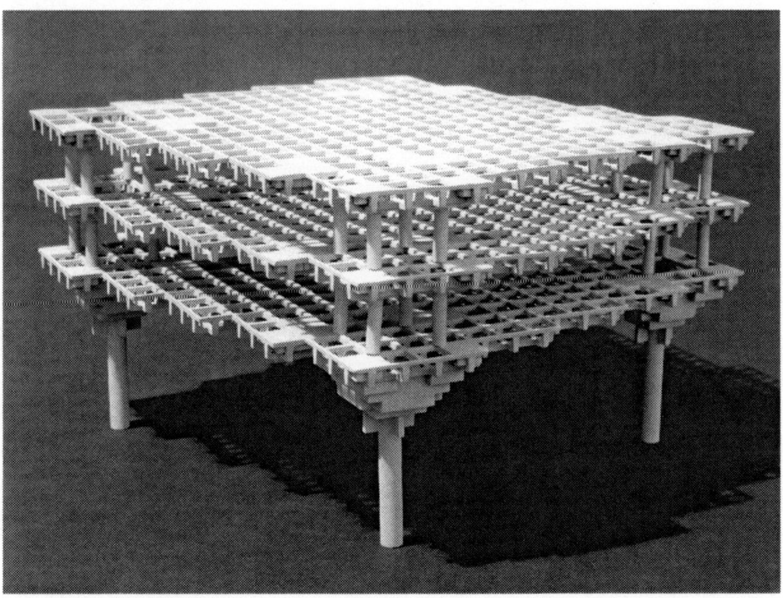

Source: University of Kentucky Center for Sustainable Cities, Richard S. Levine (architect)

have increased. A recent, flexible, concrete structural system known as the Coupled Pan Space Frame (CPSF) (Figure 10.9), which generates a complex family of building geometries, is being used as the framework for both spanning the train tracks and for creating the inner hill and urban fabric at the aboveground level (Levine, 1982; Dumreicher and Levine, 1996). The CPSF permits the negotiation of level and sloping streets on the constructed hill, giving it the sort of three-dimensional, organic character rarely seen in modern architecture and contemporary cities.

Vienna's sustainable city implantation

A train yard 1.5 km long by 200 metres wide lies behind the Westbahnhof train terminal. For many years, this yard has been an open wound in the city, dividing a neighborhood and creating near slum conditions on either side of the tracks. There have been many proposals to build over the yard, but none has offered a suitable economic proposition or an urban-architectural design acceptable to the city. Through a series of three design commissions over five years, an evolution of design proposals has emerged (Dumreicher and Levine, 1995, 1996). The alternative proposals present scenarios, which have numerous features in common and represent a development guided by feedback from various stakeholders. They all include a glazed, vaulted train shed behind the terminal building at the east end of the site. It is, in part, in the tradition of the early glass train sheds found in many major European cities, except that the glazing contains integrated solar photovoltaic (PV) collectors which deliver a substantial percentage of the implantation's energy requirements while modulating the climate and quality of light entering the terminal.

Figure 10.10. Cutaway view of Westbahnhof SUI

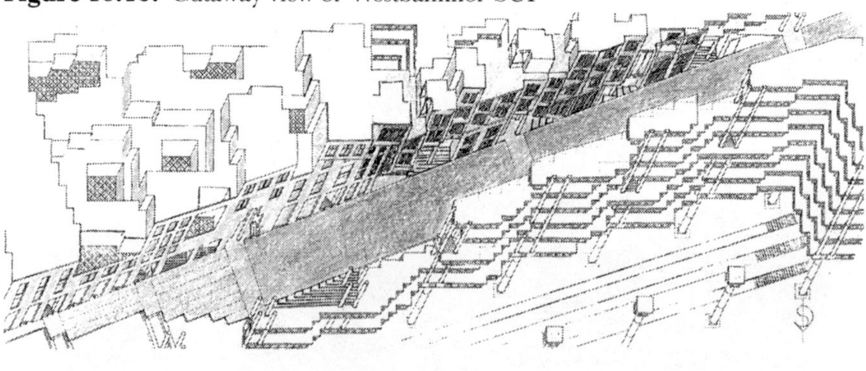

Source: Richard S. Levine (architect) and Taqi Radmard (architect and renderer)

A pedestrian street extends from the terminal and runs the length of the site to the west, parallel to the tracks, rising up into the city-as-a-hill at a gentle 6 percent slope (Figure 10.11). As the street rises, it crosses each of the horizontal floor levels, making almost every place in the city handicap accessible. At every third level (levels four, seven and ten), it passes through a public square or piazza. A streetcar runs along this otherwise pedestrian street and, after gently rising through the main piazza (*Hauptplatz*) at level ten, it descends through piazzas at levels seven, four and one to join an existing trolley track at ground level.

On the surface of the constructed hill is a human-scaled town with networks of streets, stairs, piazzas and paths weaving between three- to five-story buildings in neighborhoods of dwellings and a variety of shops and service providers. Also on the hill's surface is a winter garden growing food year round, with a network of south-facing greenhouses. At the west end, there is an east-west exchange center, which is roofed by a large terraced, ecological park and connected to the existing technology museum. There is another sloping street at this west end, which is more like a linear park,

Figure 10.11. One view of the *Hauptplatz*

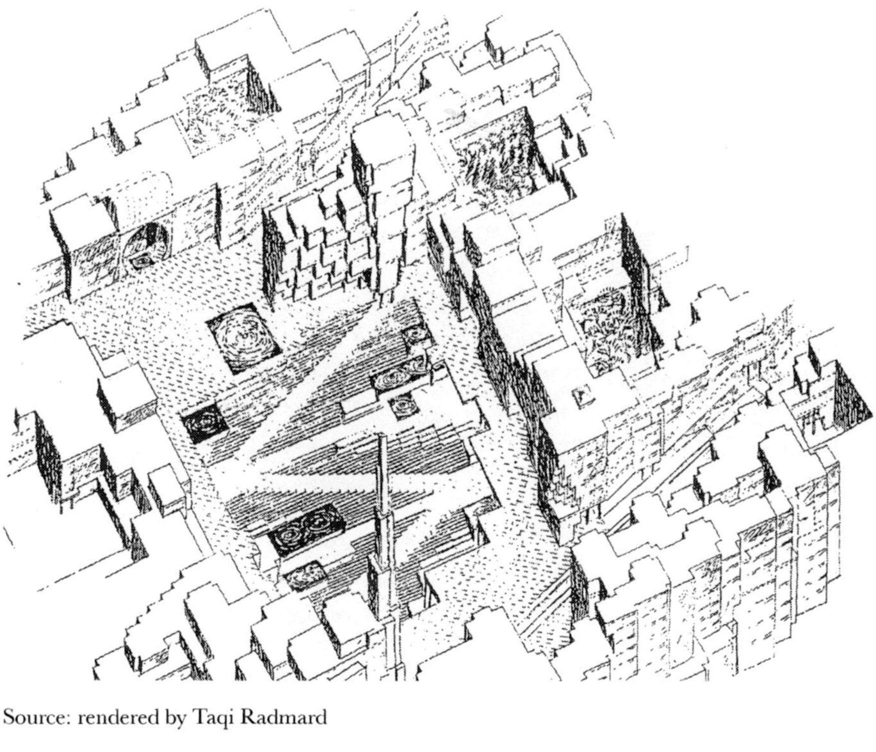

Source: rendered by Taqi Radmard

culminating at the level ten *Hauptplatz* (Figure 10.10). This main square features a water fountain, which also serves as an energy gnomon (the height of the fountain is an indicator of the rate at which the implantation exports renewable energy to the larger city of Vienna). Overflow from the fountain – supplied by rainwater collected from the site – trickles down through the linear park, feeding various ponds and other green areas along its course. The rooftops of the outer city are either glazed greenhouses or level surfaces, which are all utilized as private or semiprivate terraces, gardens, courtyards or public parks, playgrounds or piazzas.

Running inside the constructed hill, several three-story interior gallerias at levels four and seven extend almost the entire length of the site and are the major horizontal circulation levels. These inner gallerias connect to the piazzas at each end (levels four and seven), as well as to a number of courtyards to the north and south along the way. Alongside the gallerias, daylit through courtyards and light wells from above, major institutional, commercial and industrial facilities have been planned and constructed. These include infrastructure, service, parking, tracks and transportation facilities – the sort of big blank-looking structures whose overlarge scale and deadzone effects often disrupt the integrity of a traditionally urban fabric (particularly when they occur at the city surface), but which are necessary to sustain a modern urban economy. Within the interior of the city-as-a-hill they fit in well, providing maximum accessibility without compromising the human-scaled, village character.

The SUI is a totally urban construction, which multiplies value in part because it multiplies real estate. Railway services occupy almost the entire site at the original ground level, but there are many additional levels of developable real estate in the framework above, each with its own appropriate functions and activities. Because it is completely urban and has no open ground of its own, the implantation has to be linked with a rural partnerland, which is dedicated to rebalancing the energy and material flows of the city-region it supports. On this land, the agriculture and energy from solar regenerative sources would be negotiated with its urban counterpart. The implantation – together with its rural partnerland – would constitute a self-contained sustainable city-region sharing a combined SAB.

The Partnerland Principle in Practice

We have referred in this case study to the relationship between urban implantations and rural partnerlands for the exchange of goods, materials, energy, social and cultural opportunities, and other benefits. However, it is one thing to create theoretical models of how this might work and quite another to create the social and political space to enable it to happen. To create a fertile ground out of which the Westbahnhof SUI can rise, the whole 15th District of Vienna was designated as the urban region to be utilized. The region partnered with Mistlebach, a sparsely settled agricultural region in the Weinviertel to the north of Vienna near the Czech border, with approximately the same overall population figure of 70,000.

Within the 15th District, an extensive network of social, ecological and economic initiatives and enterprises already exist, but the actors were largely unaware of the importance, the value or extent of this system. Individuals often consider themselves as part of a powerless minority who at best can overcome only some of their more pressing problems. However, in the process of establishing the network, the actors come to see the relationship between their own small spheres of activity and the dynamic and increasingly powerful whole.

In activating the city-region principle, an early step was to establish partnerships between well-defined structures in the town district (Wien 15) and the region (Mistlebach), including the political, administrative, cultural, economic and agricultural institutions. The concept of a marketplace was extremely useful in establishing new opportunities. The marketplace of ideas, as well as of goods, services and cultural exchanges, becomes the generator of an expanding social network for the creation and utilization of available goods and services. Such a program was implemented, involving the farmers of Mistlebach and the grocers of the 15th district. In subsequent stages of development of the SUI, the city models and their parts become the framework for integrating an array of other systems into the city and the modeling of the ecological balances both within the city, and between the city and its rural partnerland. These systems included mechanical, electrical, material and infrastructure systems, facilities management, information, energy and material flow models, economic activity, imports and exports (input/output). The Sustainability Engine will be used to accomplish these goals. The Sustainability Engine, which is currently evolving into the Sustainable City Information Modeling (SCIM) program, is the autonomic nervous system of the SUI. During the design process, and in the governance and management of the city, the Sustainability Engine houses the energy, material flow and process models, which are studied, tested and modified in order to evaluate the progress and development of the city. During the study of the city, its processes and industries, the Sustainability Engine provides frequent feedback on the ongoing state of the system and indicates the sectors that are out of balance; this process provides issues that will be addressed in the next iteration of the Sustainable City Game.

With this design, urban sustainability promises to create the next major transformation both in architectural design and in our cities. In many ways, the sustainable city may represent the rebirth of contemporary architecture. The Athens Charter, an early manifesto of modern architecture, became a disaster for our cities. Because of the mechanistic ways in which it separated functions and activities, it reinforced economic tendencies toward unsustainability. In contrast, the sustainable city will demand a dense, diverse, highly integrated urban fabric. It will demand a new range of architectural and urban forms and structures; it will put architects and architecture at the heart of a participatory process demanding the skills and creativity of all participants. In short, it will complete the agenda of modern architecture, which never transcended its stylistic tendencies. Through the pursuit of the sustainable city, it will gain new life and value by now operating at the service of a larger societal and environmental whole.

Conclusion

The Westbahnhof project stands out as a test of the many aspects of the theory and design of sustainable city-regions. Although it was never implemented, it offers several lessons for the sustainability movement generally and sustainability advocates and practitioners, in particular. These include the following:

Urban sustainability projects must negotiate the rough waters of local-regional-national politics. Despite the positive reception of the three Westbahnhof design studies by city officials and the local-national mass media, a shift in Viennese politics and the governing political coalition resulted in the loss of its primary champion, Hannes Swoboda, and thus its political traction. When Swoboda's party was defeated at the municipal level (and Swoboda elevated to a national post), interest in tackling the longstanding and widely felt need to overbuild this central railway station and its transportation terminal dissipated. With the evaporation of political support, the wound in the heart of Vienna remained.

Popularity for public support for large-scale sustainability projects often breeds resistance. Before the political reversal, the most crucial private partner in this ambitious enterprise was the railroad company that owned the land comprising the Westbahnof railway site. It was critical that this stakeholder be persuaded to sell rights to parts of the land and allow the urban sustainability implantation to be built over the existing railroad tracks. As local negotiations dragged on for many months in the early 1990s, railroad executives increasingly sought to take advantage of their strategic position; eventually, they sought to sell off parts of the contiguous land for what they deemed greater profit. In spite of the greater sway of public institutions vis-à-vis private enterprises in the Austrian political system and its surrounding political culture, a true partnership between the railroad and the city was never successfully hammered out.

The principle of partnerland was successfully tested and proven as a key element of the theory and design of sustainable city-regions. One positive development from this stillborn effort to translate sustainability theory into sustainability practice was that the coupling of partnerland with an urban sustainability implantation demonstrated how the scale of the city-region (standing at the center of our theory of sustainable cities and serving as the putative scale for the sustainability enterprise) could be realistically implemented in real-life contexts. By working from a set of assumptions for an approximate and hypothetical calculation of the SAB of such a project, it was possible to estimate the amount of appropriated land that would be necessary to support the expected population living within the Westbahnhof implantation.

The Westbahnhof project imagined a way that metropolitan world cities could move beyond "showcase sustainability" buildings and neighborhoods. While this project design was never built, its modeling and design did demonstrate how world-class cities like Vienna could lead the way toward overcoming the recent penchant to underwrite the design and construction of urban architectural and neighborhood planning showcases of (at best) sustainability-oriented projects. Even in the 1990s, sustainability-oriented building efforts in Frankfurt, Germany and elsewhere in Europe and the United States (e.g., Moore, 2007) have largely contented themselves with using Leadership in Energy and Environmental Development (LEED) and other ecological building techniques to embed green exemplars into their urban landscapes. The triumph and tragedy of the Westbahnhof project were its boldness of vision and its practicality of design work.

Chapter 11

THE SUCCESS OF SUCCESS: THE CHINESE VILLAGE AS CATALYST OF FUTURE CHINESE SUSTAINABLE CITIES

Introduction

China is currently undergoing a program of massive, unprecedented urbanization and industrialization. Within the next five to ten years, a minimum of 200 million Chinese farmers will leave their villages to become factory workers in hundreds of new and greatly enlarged towns and cities. Accompanying this shift are strong indications of massive unsustainability from an economic, social, cultural and environmental point of view. The way in which these Chinese cities are conceived and constructed will have portentous consequences not only for China but for the rest of the planet as well (Figure 11.1).

It is vital that rapidly industrializing societies such as China's pursue sustainable alternatives to this ill-conceived development; in recent years, we have seen some institutional and governmental interest in China in a search for such alternatives. With a keen understanding of the importance of identifying pathways to mitigate unsustainable development in China, the European Commission sponsored a five-year research program addressing future prospects for sustainability in the numerous small villages in which 70 percent of the Chinese people still live.

SUCCESS is an acronym for "Sustainable User's Concepts for China Engaging Scientific Scenarios," (ICA4-CT-W2002–10007). The SUCCESS project's goal was to explore the prospects for a sustainable future for the Chinese village as a social and cultural question as well as from ecological and economic points of view. (For a recently published work generated as an outcome of this project, see: Dumreicher et al., 2005; Dumreicher and Prañdl-Zika, 2008; Dumreicher, 2006; Marschalek, 2008; Mortimer and Grant, 2008; and Shaw, Hunter and Mortimer, 2007).

Figure 11.1. Shanghai at night

Source: University of Kentucky Center for Sustainable Cities Design Studio

SUCCESS worked with seven villages in six Chinese provinces (Figure 11.2), with a strategy of initiating civil society processes to increase the quality of life, economic potential and livability of the villages through sustainability-oriented means. Teams of villagers or so-called "critical reference groups" selected by the villagers themselves worked with 40 researchers from 17 institutions in China, Europe and the United States; they considered the questions of what to maintain and what to change about the traditional village structure and way of life in the face of rapid modernization and threats of massive population loss to megacities. Keen to preserve the most supportive aspects of the traditional way of life while also benefitting from nondestructive modern improvements to that way of life, participants were eager to explore structured means of researching a sustainable future. Village teams created multiple alternative scenarios of possible sustainability-driven futures for the villages.

The Village as Generator

The traditional Chinese village has gradually evolved over thousands of years; it represents a slow-to-change, balance-seeking adjustment between cultural and material patterns and the natural landscape (Figure 11.3). Over the course of time, different cultural and architectural typologies have sprung up and crystallized in these villages, each unique to its geographic location and cultural circumstances.

Figure 11.2. Chinese map of villages studied by SUCCESS

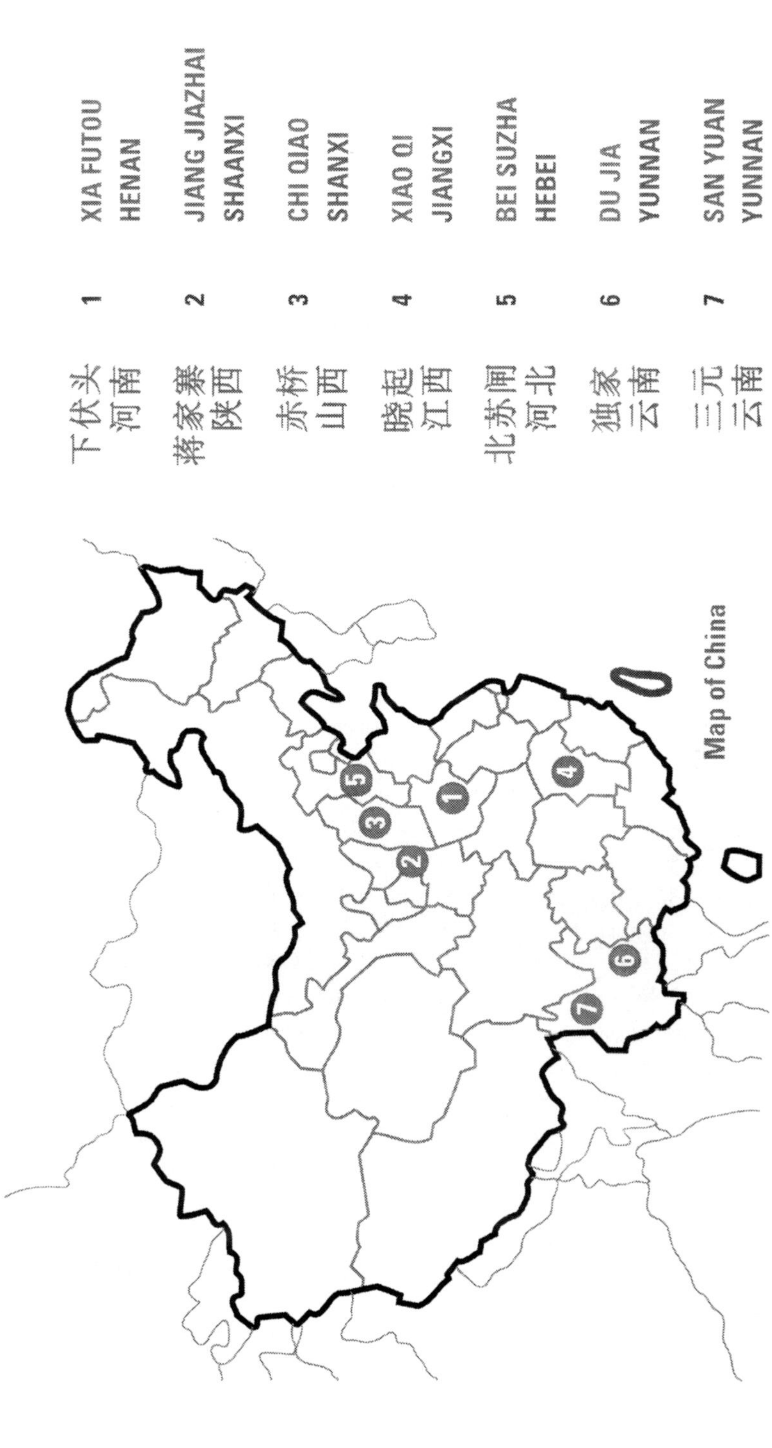

Source: Heidi Dumreicher, Oikodrom, SUCCESS Project

Figure 11.3. Rural village in Yunnan Province, China

Source: University of Kentucky Center for Sustainable Cities and Oikodrom, SUCCESS Project

With the rise of the People's Republic of China in 1949, particularly during the ten years of the Cultural Revolution beginning in 1966, the villages were subject to wrenching revolutionary changes when compared to the slow preindustrial, traditional evolutionary process that had brought them up to the mid-twentieth century. As a result, their local culture, architecture, and most importantly, their "protosustainability" continue to be threatened by rapid and uncontrolled modernization. We use the term "protosustainability" to describe a steady-state condition of early towns and villages around the world, where the metabolism of the town is carried out as a balance-seeking relationship between the local culture and the natural environment but where social and life-quality issues are far from modern expectations. For example, the introduction of electricity and its infrastructure was fairly rapid, whereas the integration into the traditional life patterns and village architecture has moved at a much slower pace.

The generations-old craft traditions that have created a complex, beautifully crafted, functional system of building that coevolved with the local culture illustrate this phenomenon. Today, electric wiring and plumbing are still crudely slung between beams and columns in stark contrast to the elegance of the rest of the structures. This is an example (in microcosm) of the awkward unsustainability that is happening on a larger scale within the cities of China. Through their rapid urban and industrial expansion, the Chinese have not had sufficient time to adjust their social traditions, urban forms and technological base – indeed, their modern evolution has been almost totally devoid of cultural considerations – and such adjustments to human need might lead one to the conclusion that it might not even be possible. This new urban metabolism and its interruption of indigenous cultural patterns has evolved new, unsustainable conditions that will be increasingly difficult to correct. These economically booming cities are overflowing uncontrollably onto the traditional Chinese landscape, damaging the structures of its formerly protosustainable lineage.

The Chinese village, unlike the Chinese metropolis and despite a few modern influences that have lately crept into the system, is still largely protosustainable. This means that for the most part, in a typically unmonetized and locally self-sufficient economy, Chinese villages still produce their own food, manage their own waste, feel the effects of their own problems and manage them all locally. For this reason, the SUCCESS project identified the traditional village as an appropriate place to initiate research into the implementation of sustainability processes for the future Chinese sustainable town. That is to say, it was an assumption of the SUCCESS project that an understanding of village life and its metabolism might be able to provide the beginnings of a model for future Chinese sustainable development.

In modeling village metabolism, we first eliminate the few unsustainable traits of the existing village system and replace them with their sustainability-oriented counterparts. This renders it possible to project the evolution of future, much larger towns from the nodes that are already working with sustainability-oriented materials through sustainability-driven processes. In this multiple alternative scenario-building process, little by little we enlarge the village model using only sustainability-oriented means until the overall system is no longer a village, but a full-service town with all the businesses, institutions and life opportunities that would make it a desirable place to live.

The 800,000 Chinese villages where most Chinese people live are moving toward extinction for a variety of reasons. First, government policy, economic pressures of mass industrialization and urbanization, along with the growing influence of consumerist culture, are moving villagers to the urban centers in droves. Second, Chinese children are being educated for a life, and for that life's opportunities, that cannot exist in a small rural village. Third, realizing that in the West only 2 to 3 percent of the people are farmers, the Chinese government is intent on realizing a successful future approaching these numbers – something that officials apparently believe cannot be done with the traditional lifestyle and farming methods to be found in the villages today.

Methodology

As a research project (not a development project), tangible results of development implementation schemes were not part of SUCCESS's purview or expected outcomes. However, as sustainability demands engagement and participation, such future scenario building without an agency for pursuing the chosen scenarios would not only be an empty exercise, but would be damaging to the spirit of the villagers whose hopes would be raised only to be dashed by the hard reality of nonperformance. We faced this dilemma head on within the design of the SUCCESS project. A crucial part of SUCCESS was the regular engagement of teams of villagers – the project's local experts – who met regularly as the project's "critical reference groups." We considered the villagers a vital part of the research team, along with the Chinese experts and foreign experts, and paid them a modest amount for their time and work (elevating the villagers well above mere subjects to be studied). There were in effect two sorts of experts involved in the SUCCESS project – the local expertise of Chinese villagers and the disciplinary, professional experts from China and abroad. There were outside experts in many different disciplines. They included a number of participants and stakeholders who were representatives of various Chinese agencies, universities, businesses and officials at a number of different levels of Chinese government.

Among the researchers, the group sorted itself out as cohorts of social scientists, natural scientists and a few architects and urbanists; social scientists played the dominant role in group facilitation. Although all the researchers from a large variety of disciplines were chosen for the project because of their prior expertise in "sustainability," it turned out that sustainability was understood as many different things according to the diverse perspectives of the different disciplines. Though experts in each discipline understood sustainability to be an interdisciplinary endeavor, the group encountered some difficulties in addressing sustainability within a transdisciplinary framework. Indeed, the fractionalization of the group along lines of disciplinary awareness occurred in large part because of the funding structure under which such collaborative research was organized.

The terms of the research grant that supported SUCCESS (ICA-A4-CT-2002-10007) divided the project into a large number of "work-packages," each with its own unique listing of "deliverables." Nevertheless, at the heart of the grant proposal, the project leaders located a number of integrative work-packages explicitly for the purpose of requiring researchers to invest time and energy in coming together to synthesize alternative sustainable future scenarios for each of the villages. We based the project deliverables upon the findings of the individual work-packages, with major input from the villagers themselves who contributed through their critical reference group and the social science teams of Chinese and European researchers. In fact, by revealing the details and the connections of their village structures, the villagers were the principal source of integrated content for the SUCCESS project.

The integration sessions, held in the villages themselves, were long, intense, contentious, and ... productive. The ebb and flow of ideas and debate shifted from the specific disciplinary determinations up through the larger-scaled village production and social patterns, and back down again to the individual disciplines. The natural scientists tended to be the holders of the most detailed technical information, while the social scientists (using a variety of sophisticated tools) were able to work with the local experts. The social scientists were able to ferret out which ingrained patterns of life and work were essential parts of a living culture versus those that were a holdover from former times that held no permanent value. In many cases, someone had to make the determination whether the "holdovers" were detrimental to current life patterns and to any sort of progress or whether they were new manifestations of a Chinese urban culture just starting to migrate to these rural villages. The natural scientists tended to see sustainability as a collection of essentially quantitative technical issues with social implications, and were content in their disciplinary rigor to shift such implications into someone else's domain.

The social scientists, for their part, saw sustainability as a social question with some technical details for others to work out. These differing worldviews posed significant consequences for true collaboration. The natural scientists viewed the work of the social scientists as not being very scientific, while the social scientists tended to view the work of the natural scientists as not being very important. This dynamic affected the work of the natural scientists the most. When they found it hard to deal in an integrated way with the entire team, it became all too convenient for the natural scientists to retreat to the comfort of their individual work packages and deliverables, and as a result, their detailed quantitative findings would become all the more difficult to combine with the findings of other disciplines.

This inherent tension was somewhat mitigated by the understanding that sustainability of necessity transcended disciplinary boundaries. To some extent, the influence of the architects and urban designers moderated this tension, but we put that tension to good advantage primarily through the leadership of the social scientists whose disciplinary understanding allowed them to deal with the village, its life, its metabolism and its future at the scale of the whole. This was possible because of the ongoing participation of the villagers through their critical reference groups and their continuing development of the village civil society process.

We can pursue this strategy of dealing with the simplest whole in an effective manner by going back to one of our first principles of sustainability. Rather than taking the value-based disciplinary quarrels at face value, the team was able to unite around the grand possibilities created by a kind of negotiation process among the diverse team of experts. We call this negotiation the Sustainable City Game (see Chapter 6). The negotiated rules and boundaries within the game create a safe place for conflict and that "conflict…is an important source of creativity in the sustainable city" (see the "Five Operating Principles" delineated in Chapter 1).

Although SUCCESS was a research project and not a development project, we were ultimately able to dedicate a small amount of the research funding to one small development project in each of the villages. This made it possible for the critical reference group to consider alternative scenarios that would actually be implemented and made them stakeholders in actual (albeit limited) sustainability-oriented projects. In these poor villages where the per capita cash income was in the neighborhood of $200 to $400 per year, there was no shortage of projects on which to spend funds. The immediate preferences of the villagers did not always fit with the researchers' ideas of a long-term sustainability agenda. The villagers tended to see the opportunity of receiving outside cash as a way of paving roads, buying computers for the local school or repairing water systems. All of these development needs would be easy improvements to make, but were not necessarily ones that would establish a longer-term momentum toward sustainability. On the other hand, perhaps the strongest evidence of success from the researchers' points of view would be the establishment of local civil society processes conducted and managed entirely by the villagers themselves that would outlive the few short years of the project. Projects favored by the researchers were actions that established sustainability principles in the minds and common actions of the villagers, and that would set in motion and synergize local resources and sustainability-oriented patterns that were already in place.

Success – Some Examples

It may serve us well to describe a few of the particular outcomes of the SUCCESS project, particularly those that constituted actual development. One phenomenon we observed over and over again was that "nothing succeeds like success." Just the fact that a large group of Chinese and foreign researchers chose to converge on small, seemingly insignificant rural villages made local and regional officials take note, often with action. When we first visited Beisuzha village in Hebei Provence, we discovered that all the previously muddy streets had been paved with asphalt just in time for our visit – something we would not have favored, at least not before engaging in a protracted stakeholder process. Although this village had developed a robust civil society process, it tended to treat the small project funding more as a windfall. Villagers spent their discussions in arguing about how they should divide the money, instead of constructing scenarios where the initial actions would have longer lasting consequences for the village. Nevertheless, they made good use of the funds by providing computers for the local school, but this was not a use likely to give the civil society meetings an extended life.

In Dujia, the smallest and most remote of our villages, the local authorities made funding available to the villagers in addition to the SUCCESS small project funds. Materials were donated for constructing a biogas digester for each household above which pigsties were constructed to deliver cooking and lighting gas, as well as supplying each household with a solar hot water heater. These simple and inexpensive actions along with a modicum of hydro-powered electricity were enough to supply all the energy needs of this, the poorest village (Figure 11.4).

In Xia Fu Tou, in Henan Province, the stakeholder process generated the decision to convert an abandoned traditional dwelling in the poor part of the village into a solar-powered bathhouse, with the villagers playing a major role

Figure 11.4. Biogas digester in Dujia

Source: Richard S. Levine and Oikodrom, SUCCESS Project

in negotiating its design. Although we carried out the scenario-building process that resulted in this decision through the village critical reference group process, research partners at Zhengzhou University completed the actual design of the facility. Although the researchers had created elaborate models and three-dimensional drawings of the small project, the villagers were unfamiliar with these methods of considering a new project and seemed uneasy about the proposal by outsiders. One of the breakthroughs occurred when we went back to the negotiation process with the villagers as the major stakeholders. In the courtyard space, adjacent to the building that the villagers would convert to the bathhouse, we laid out a plan of the building in the dirt using stones and broken roof tiles to create the configuration of the proposed project. The villagers then walked through this virtual bathhouse, discussed their reactions to the proposal and started to rearrange the walls and spaces. They would walk next door to the actual building to imagine how their evolving scheme would work within the existing building. The villagers discussed the alternative plans that were proposed and they quickly reached a consensus.

The completed solar-powered bathhouse was a beautiful facility built by the villagers and was a big success. It was not only a building. Throughout the process, the villagers developed such things as the management, operation and maintenance of the bathhouse (continuing the same civil society process). (See Figure 11.5.)

The villagers loved it and used it. But the presence of the bathhouse caused an imbalance in the previous stable village condition. There were two parts of the Xia Fu Touo village. On the hill stood the oldest stone house in what was now the poorest part of the village. In the valley were regular rows of communist-era concrete houses where the more well-to-do villagers lived. The wealthier people largely ignored the SUCCESS project, seeing it as a minor upgrade to keep the poor inhabitants from complaining – that is, until the project succeeded. When the entire village realized what a wonderful project had been built on the hill, the wealthiest families down the hill appropriated the bathhouse for themselves, locking the hill dwellers out of their own project. With this action, the hill dwellers vandalized their own project and made it unusable – and so it remains today. In Sha Fu Tou, a project that was intended to bring the villagers together to create a civil society process where they could negotiate their mutual sustainable future resulted in just the opposite.

In San Yuan village, in northwestern Yunnan Province, in a minority ethnic area of Naxi peoples whose native language and religion was undergoing a revival, an extended participatory process generated the decision to build a bathhouse for women. Although the scenario-building process was almost entirely in the hands of the local villagers, at the end, a local village leader hastily diverted these funds to another project. It seems that at a late stage – in some

Figure 11.5. Renovated house in Xia Fu Tou

Source: Richard S. Levine and Oikodrom, SUCCESS Project

form or fashion that would not be revealed to the researchers – the proposal had violated some local taboo, or perhaps in the absence of oversight, political momentum shifted towards the desires of a vocal minority. Nevertheless, the local projects were only a small part of the SUCCESS project.

In San Yuen, with the revival of the local religion and customs that had been suppressed and almost disappeared during the excesses of the Cultural Revolution, local women were empowered as never before through the influence of the civil society process and new forms of "tradition" emerged.

In Jiangjaizhai village within in Shaanxi Province, villagers created a model agricultural village, where the introduction of Western milk cows fostered entirely new patterns. Jaingjaizhai serves as a powerful example of unintended consequences resulting from sustainability-oriented actions that developed too late in the scenario-building process. Without a previous scenario-building process to examine new opportunities and problems arising from the introduction of a dairy operation, dirt streets became almost impassibly muddy with huge piles of (somehow) unanticipated cow manure accumulating as an environmental problem instead of being used as an opportunity.

Although the future scenario-building exercise was successful, it may have been too abstract for the situation as it related to a far too distant future; and that future was not easier for the villagers to relate to their current problems and opportunities. As a result, the available funding was diverted to the pressing and immediate problem of paving the streets instead of looking at the dairy operation more holistically and perhaps putting the manure through a biodigestion process, creating biogas for home cooking and using the residue as a fertilizer and soil conditioner. In contrast, paving the streets prevented the future installation of pipes that could take the biogas so generated to each of the houses.

During the time of the project, villagers built a new community hall, but no one ever introduced the idea into the scenario-building process as a need or possibility. The new hall was built in a primitive, box-like clumsy style and located outside the village. Imagine if villagers had put the hall in the heart of the village; it could have been a catalyst for augmenting the civil society process (which clearly had not been successful in the past).

The pace of Chinese development is rapid, and the speed at which decisions are made is illustrative of the sort of pitfalls that are likely to occur when the civil society process remains largely disconnected from the actual development process. Rapid development in the modern Chinese fashion is done with insufficient consideration for – or an understanding of – the consequences. Unfortunately, visits by the entire research team were too infrequent to keep up with the changes as they were occurring (even though research partners were located in nearby towns).

In another village, Xiaoqi in Jiangxi Province (perhaps the most visually beautiful village blessed with vibrant local participation), one of our professional Chinese partners charged with local development chose to short circuit the scenario-building process and use the small project funds to build a much-needed water pumping facility even before the SUCCESS process was initiated. However, the pumping structure was inappropriate because of its violation of the elegant and consistent materials, proportions and construction methods of the village; and it manifested itself as an interruption of the village scale and spatiality. The pity was that Xiaoqu had one of the most active participatory stakeholder scenario-building processes of all the villages despite the lack of additional resources. The scenario-building process was hindered when no further local project could arise from the process. This in turn weakened the prospect that the civil society process would continue beyond the time span of the project.

The Scenario-building Process

The Chinese village is a visual model of its own metabolism. That is to say, a largely unmonetized economy in the village renders visible in each family courtyard all the materials and implements, animals, furnishings and structures that are used in the daily, seasonally and yearly cycles of life of the village. It therefore becomes a relatively simple matter to aggregate this observable information, together with good data of its magnitudes, and create a metabolism model of the village. Such a model also includes the land area, the agricultural activities and other resources employed by the villagers, the time spent on the various activities, and the inputs and outputs from outside the village. Because of the largely local economy, these inputs and outputs are relatively easy to account for. It is this information presented in accessible diagram form that permits the construction of a system model of the village metabolism.

We created the first models using a rather intuitive systems modeling program called STELLA (Systems Thinking Experimental Learning Laboratory with Animation), a popular tool offering a practical way to dynamically visualize and communicate how complex systems and ideas operate. Looking to the future, a subsequent (unfunded) proposal was to move forward with the basic metabolism model and then regress that model by eliminating the small amount of modern, unsustainability-based aspects which commonly show up in the village economy. These inherently unsustainable activities included the use of coal briquettes for cooking, gasoline to fuel the occasional moped, small amounts of agricultural chemicals, etc. When engaged in a scenario-modeling process, it is relatively quick and easy to replace them with their sustainability-oriented counterparts.

One of the recommendations of the SUCCESS final report (Dumreicher, 2006) was to develop the possibility of "saving" many of the villages by enlarging them to the size of a full-service town (in China a "town" can be up to a million inhabitants). In such a town, a great diversity of life opportunities can exist and a sufficient variety of economic activities, including farming and industry, can provide the resources to balance the urban metabolism within the framework of a Sustainable Area Budget (SAB). Such a town would become the prototype for the Chinese sustainable city. We will develop the urban model through the Sustainable Village Game using the Sustainability Engine or Sustainable City Information Modeling (SCIM) software. Modeling will begin with the sustainability adjusted village metabolism systems model and, through an iterative participatory process, grow the village model to its mature size using sustainability-oriented means, methods and materials. Through this enlargement process, we will develop a variety of alternative town scenarios depending upon which direction the different teams of stakeholders and experts take the process. Experts include villager inhabitants and outside experts, specialists, technicians, architects and urban designers. Each participant will work from the perspective of her or his various self-interests, ideas and creativity. Through successive iterations, participants will fold the best ideas from the competing scenarios into successive proposals, so that participants can bring forward surviving scenarios, each becoming more complex, more responsive to competing interests and more robust than any one proposal alone. The engine will tend to keep the process on track – that is to say, participants will be mindful as to how their initiatives will affect the balance of the town-system and will play the sustainability game accordingly. The emerging scenarios will be treated as architectural/urbanistic and countryside-scale models, as well as systems models inclusive of partial models of local places and activities.

The alternative sustainable scenario-building process takes a good deal of its structure from the architectural design process and, at the same time, employs both scientific analytical methods and systems modeling feedback tools. The current metabolism of the villages is studied and modeled, and unsustainable practices (i.e., the use of fossil fuels, agricultural chemicals, unsound agricultural practices, etc.) are replaced in the model with sustainability-oriented equivalents. Participants will develop future alternatives as systems models of present and future scenarios. They will pose "what if" questions and model different scenarios to assess both their balance as well as their performance in the village-as-a-system. Each scenario is balanced within its natural budget – its SAB – assuring that any possible scenario that is eventually pursued satisfies the multilayered criteria of sustainability.

With the experience of modeling the Chinese village and projecting it into the future on a sustainable basis having been established, the Chinese sustainability-driven village is used as the starting point for larger towns and cities whose growth and development proceeds through this same sustainable scenario-building model to develop diverse, vibrant new towns for China's future.

Sustainability as a Systematic Balancing Process

The proposal for implementing a Sustainable City Development Process as outlined here is amply supported by the experience of the SUCCESS project (research published in Dumreicher and Prañdl-Zika, 2008). It is important to note that key systemic underpinnings of sustainability do not currently underlie the dominant sustainable development initiatives in China. Though the dominant paradigm is amenable to a number of "green" efforts, it fundamentally lacks a systemic engagement of the town-system as a complex living, social and material metabolism. If, in the future, the Chinese village – evolved as a modern, developed town – were to incorporate a collection of widely accepted sustainability strategies and actions, and as a result, create scattered individual "green" projects, this in itself would not be a demonstration of sustainability. In fact, it is impossible on a case-by-case basis to establish particular standards that, in the aggregate, would constitute sustainability.

The critical point here is that the sustainable town will not be simply a collection of sustainability-oriented components. This is so because a sustainable city as a complex adaptive system depends upon an ongoing rebalancing governance and urban design process that functions only when developed as a *whole* system. For instance, it would not be useful in a sustainability regime to establish criteria for maximum CO_2 emissions for a given factory, or even a group of factories. Rather, a completely sustainable town-region would have to balance out the CO_2 emissions within its SAB region. If it chose to allocate a large part of that budget to a particular factory (due to its importance in the sustainable metabolism of the town-system), this would be perfectly acceptable as long as the overall budget of the town-system would not be exceeded, and if, on a net basis, excess CO_2 would not be exported beyond the town-region or into the future.

It would be most unfortunate if the industrialization of the Chinese protosustainable villages proceeded based upon the "best environmental practices" of the present. This is because, as environmental standards are continually raised, many of the best practices of the present will become unacceptable practices in the future. On the other hand, a town-region that has negotiated its urban processes within its ecological budget (its SAB) cannot become obsolete in the future unless it abandons the ongoing sustainability balancing process of urban governance.

The sustainable Chinese town of the future is certain to become a highly desirable location for industry, tourism, investment and living. The terms and circumstances by which development can occur must be kept carefully under control. The traditional Chinese village is an excellent starting point for the application of such a sustainability process as described herein.

The Sustainable Town-Region: Metropolis vs Village

The sustainable town-region (STR) is a new urban model that is particularly well-suited for Chinese development conditions. This model is of a very dense urban design for a small town that incorporates the diversity of economic activities, occupations, services and industries that are both necessary and desirable in the small sustainable town of the future. It is small enough to be on a walkable pedestrian scale, and thus has no need for private vehicles within the town and would require few vehicles of any sort. A large agricultural area – corresponding to its population-based SAB – will surround the town. This area is large enough to supply all of the land-based resources on a net basis (energy, food, water, etc.) to sustain its industry and way of life. The town is small enough to be affordable to build, and can be constructed in a short period. With its success, it will provide the momentum and enthusiasm for the building of more and more sustainable town-regions, each with somewhat different activities and industries. A network of such STRs should be well linked by public transportation. The projects will be an example of both sustainability principles and processes, as well as a demonstration of a desirable way to live.

This process of development will be much more amenable to mirroring the many favorable aspects of traditional Chinese urban and residential patterns than are the many foreign-influenced, unsustainable urban developments now emerging all over China. The new urban model that we have developed for China is small enough to be walkable, yet large enough to have a full range of urban services and opportunities. In addition, the model is dense enough to maximize urban qualities and amenities, all surrounded by its SAB-determined land area to provide land-based resources on a net basis (food, material, energy, etc) in order to sustain the resulting human-scaled town.

The sustainable town is conceived as a "learning ecology" (a phrase coined by Kevin Lynch) that learns from its successes as well as its mistakes. It avoids the American illness of urban sprawl. It also avoids the Western reliance on the automobile, which is a force that exacerbates every other urban problem. Our analysis clearly demonstrates that a rapid increase of private ownership of automobiles in China will quickly clog existing cities and make the development of new sustainable cities impossible. The sustainable town-region is the smallest significant project that can stand on its own as the minimum unit of sustainable development, neither requiring the car nor polluting the natural resource base upon which it relies.

Implementing the Sustainable Town Process

After having selected the protosustainable traditional Chinese village where the sustainable town project is to be developed, it is crucial to get the process right. The process will be carried out in conjunction with the operational definition of sustainability as we have developed it, as it is largely embedded in the European Charter of Cities and Towns Towards Sustainability (Aalborg Charter) and as developed further by the Center for Sustainable Cities in various other projects. In China, sustainable development will take on a very different form than it would in Europe under the Aalborg Charter. The reason for this is that Europe is a continent of mature towns and cities, and development must inherently mean gradual change within these established towns and cities. In China, studies (Roberts, 2008; Logan, 2008; and Chaoui, Leman and Rufei, 2009) have projected an incipient migration unparalleled in human history. This will necessitate construction of numerous new towns and major expansions of existing towns and villages. Thus, there is the possibility, and arguably the necessity, for new types of urban structures as well as new kinds of urban governance and management. A vast array of new choices will open up that are not possible in either Europe or the United States.

The sustainability process in Europe relies on the presence of participatory, bottom-up citizen-stakeholder activity. The kinds of processes for gradual change that are being institutionalized in Europe according to the principles of the Aalborg Charter make great use of such citizen participation. With major new developments such as hundreds of new towns cropping up within the span of just a few years in China, processes with strong citizen participation and bottom-up character are not likely to gain political traction or be economically feasible. This carries some inherent disadvantages, but also offers some significant advantages if the people and institutions that will shape the towns are chosen in an appropriate manner.

In shaping a new or greatly transformed village or town to operate according to sustainability principles, a team of experts, technicians, potential or surrogate stakeholders and citizens will assemble to play the "Sustainable Town Game." Initially, the researchers and consultants will set the town game up as a systems modeling process at the town-region scale. The information database is filled with documents and embedded with relevant data on local characteristics relating to economy, environment, social and cultural conditions.

The team of participant-stakeholders will include experts from many different disciplines, but start with experts who at the same time have a strong background in sustainability principles as well as in collaborative efforts and participatory processes. It will also include systems model builders and technicians as well as architects and urban designers skilled in working with sustainability-oriented concepts. From this point, the participants will work with a variety of town-region scenarios. Though the stakeholder-participants or the experts and technicians may prepare the scenarios beforehand, the scenarios can also arise out of interaction and collaboration between the different participants during the "game" process itself. After the team designs or negotiates three or more of these scenarios, the scenarios are tested and evaluated through a systems analysis using the Sustainability Engine.

The feedback obtained in this process will give a solid indication of the ways in which the different aspects of the town-region scenario are working well; it will also reveal the aspects that are far from being in balance within the system. In the early iterations of this process, we expect that many aspects of these scenarios will be far from their balance points. Since the main criterion for sustainability is that a sustainable town-region operate near the balance points of the system or approach those balances, revealing far-from-balance conditions is an indication of which issues must be addressed in the next iteration of this scenario-building sustainability game. After participants obtain feedback from the engine, they take stock in contemplating subsequent moves and modifying the scenarios. New ideas and suggestions can come from any quarter. When a number of different scenarios are in play, it also becomes clear to the participants that each of these scenarios has strengths as well as weaknesses.

Playing the Sustainable Town Game becomes a process of working with the strengths of the competing scenarios, while at the same time trying to eliminate their weaknesses. After a number of iterations of this process, each of the surviving scenarios becomes more complex, more comprehensive and more robust. At the same time, diverse stakeholder interests must be incorporated into each of these scenarios. Through this process, each of the participants gains a sense of ownership of both the process as well as each of the emergent scenarios. The sustainable town, first as a game (and later as an actual town to be constructed and managed through a continuation of this stakeholder process), gains the confidence, support and emotional ownership of all of the participants.

Because of the nature of the SUCCESS project, only the first tentative steps of the participatory process outlined above were possible. Even though such processes were quite alien to the rural Chinese culture, in each instance the process yielded at least some measure of success and the new tradition of participatory civil society process was established. In terms of its research agenda, we should regard SUCCESS as a substantial success, yet the outcomes were at best small steps in terms of tangible results. In an actual city-building, sustainable city-region development process – given the high risks involved – the integrity and the viability of the process must be assured and protected. This is difficult in a process like the one outlined here, where both top-down and bottom-up aspects are in dynamic interaction. We believe that the sequential unfolding of the sustainable city game is an excellent framework for building confidence in the process, while at the same time developing and projecting more and more detailed and validated sustainable town-region models that elicit an emotional co-ownership of all stakeholder participants. As they buy into both the process and the emerging city design, the city attracts a strong constituency of supporters. At the same time, a strong virtual community develops in advance of the actual community that will emerge in the construction and the inhabitation of the town.

Systems Modeling with the Sustainability Engine

The feedback tool in this alternative scenario-building process is a systems modeling tool that we call the Sustainability Engine. The engine is, as we noted in an earlier chapter, a software system that is still under development; in its early form, it was based on the STELLA software. At present, it is a simple mathematical model into which data from the alternative scenarios is plotted on a systems diagram of each scenario. Users run the system under a variety of "what if" scenarios in an effort to balance as well as to maximize the performance of the town system. In its current version, we are porting the engine into a Building Information Modeling (BIM) platform, which we are calling Sustainable City Information Modeling (SCIM). In this version of the engine, we have constructed three-dimensional city models and partial models (or modules) to house the balancing information developed in the different scenarios, and a much more user-friendly interface is used. In the advanced version now under development, diagrams of the village system are associated with plans and three-dimensional images of the alternative town scenarios. Thus, ordinary townspeople (with the assistance of architects and technicians) will find it easier to enter their ideas and preferences into the "what if" scenario modeling process; they will be able to view images of the urban consequences and get feedback as to any quantitative performance data.

In the SUCCESS project, we developed such a systems modeling process for Chinese villages. The Chinese village is, of course, smaller and simpler than a modern town, but we can model its system characteristics in much the same way. For demonstration purposes, some very simple systems models of the Dujia village in Yunnan province were developed. We demonstrate the modeling process in several sections below. In the first sections, we explain how to make a systems model and how a systems model works. Systems diagrams that use "intelligent" icons to typify the relationships between the different parts of the village system lie at the heart of these models. Because these mathematically based systems icons represent the actual metabolism – that is the energy, time and material flows in a village – it is a simple matter to add or subtract functions from the village or to change their relative quantities. By doing so, we can ask "what if" questions about the village system. As a demonstration of this, in the last section of this chapter, we take a small part of the agricultural economy of Dujia (the growth of certain commodity crops for export income) and experiment with changing the amount of agricultural allotment devoted to the different crops.

In conducting these simple "what if" experiments, we made an unexpected discovery in the Dujia model. The production of sugar brought in a substantial percentage of the yearly income of the village; however, when we modeled this process, we discovered that it had an associated negative net cash flow. The villagers would actually be able to eliminate almost half of their annual labor and still have more cash on hand if they did not plant at all – a significant fact that was not known to the villagers. Although the sugar cane harvest generated a large amount of income at the two yearly harvest times, production expenses accumulated gradually throughout the year and were not accounted for separately from the other agricultural expenses. Of course, this example is a very simple one. Researchers can make many more complex determinations by asking numerous "what if" questions throughout this process. Through these alternative "what if" models, we can explore the consequences of not planting sugar cane, like using the now-available land to plant other crops. Likewise, villagers could use the time gained for other activities such as using produce to create a value-added product instead of selling the agricultural product directly into the commodities market at a very low price. The irony of modeling the elimination of the sugar cane crop is that the cane is not a traditional crop; it was only recently introduced into the village agricultural regime. Much of it has been (illegally) planted on steep hillsides, which is starting to cause severe erosion, contributing to the silting and flooding of the river downstream and degrading the longer-term prospects for the village.

Figure 11.6. Sugar cane being harvested in Dujia

Source: Richard S. Levine and Oikodrom, SUCCESS Project

Figure 11.7. STELLA diagram of Dujia

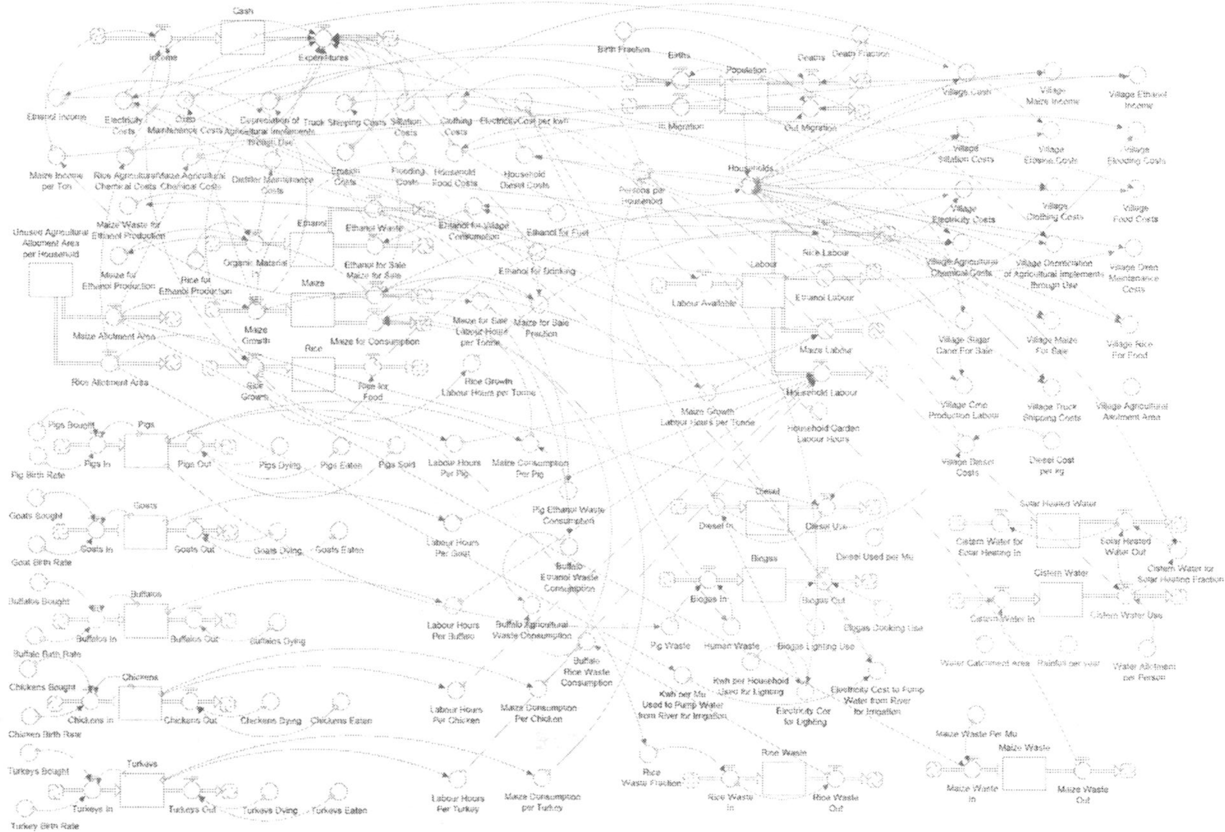

Source: Michael Hughes and Casey R. Mather, Center for Sustainable Cities Design Studio/SUCCESS Project

We can model all of this and respond accordingly as the village expands to become a larger sustainable town. (See Figures 11.6 and 11.7.)

A key component in building any systems model is that we carefully define the boundaries of the system. The sustainability modeling process has two kinds of boundaries. The first boundary is constrained by the conditions of the actual town: the physical, geographical, economic, climactic, soil and water limits, as well as the cultural, social, administrative and legal limits that may exist. The second type of boundary is what we have designated as the Sustainable Area Budget (SAB), which is similar to the Appropriated Environmental Space in Europe (see Chapter 5). The SAB concept makes it possible to use the footprint concept, which is an analytical (reductive) approach in a design process (a synthetic process). The SAB concept establishes the land and resource budget, which we can then assign to a given town-region.

The concept develops what others have called the "fair Earthshare" (Wackernagel and Rees, 1998) of the world's renewable resources. In principle, this means that each individual is entitled to a six-billionth of the Earth's bounty on a regenerative basis. Thus, a town would be entitled to the aggregated "fair Earthshare" of each of its inhabitants, interpreted as resource-based and biologically productive land area. While this principle is simple enough, there is a lot to be decided in terms of the *exact* area to which each individual is entitled. There are many variables that determine usable land, including fertility, water, climate and so on, because not all land is of equal quality. The presence of nonuse value of wildlands – as the biological basis of biodiversity and ecological productivity but not direct human use – is also of significant and growing concern. Though there are possible variations in methodology, the amount of land accruing to each individual by any of the different schemes of calculation, with their different underlying assumptions, does not vary a great deal.

This means that, for China's purposes, we can establish a fairly consistent land budget by dividing China's land area, which is 9,326,410 sq. km, by its population of 1.3 billion people. This gives an SAB per person of approximately 7.2 *mu* of which 15.4 percent or 1.1 *mu* would be arable land. (The *mu* is the Chinese measure of land area and is equivalent to approximately .1647 acres.) Although we may wish to modify the determination of this number in the future for a variety of reasons, for now it should be a satisfactory basis for the determination of the SAB for any prospective sustainable Chinese town. (Ultimately the SAB needs to be accorded a global meaning, to be negotiated by international agreement.)

The sustainable town (particularly in the early stages of its emergence) would be a town that maintains the traditional Chinese strong agricultural base. Thus, designers should plan and calculate for the town to provide, among other things, all of its own food and energy requirements on a net basis. In most cases, a town with a strong agricultural base is likely to produce a large surplus of agricultural products, which in turn can be used to supply larger cities and serve as a source of some of its export income. This export activity will actually increase its SAB allotment because it will be provisioning the larger cities that are unable to provide for all of their own needs. Therefore, the sustainable town or city will effectively appropriate some of the larger cities' SAB. This will add to the smaller Chinese cities' economic and sustainability potential and value – a land ethic which in a rapidly industrializing society is increasingly being taken completely for granted. (For additional treatment of the systems dynamic model and its application, see Levine et al., 2008.)

Conclusion

In recent years the trajectory of Chinese urban, industrial and economic development has been nothing short of phenomenal. As this trajectory continues there are strong signs that while quantitative growth remains extremely robust, it will not be possible for such a path to continue indefinitely. Its path will increasingly run up against environmental limits, social and cultural limits and even economic limits. It will be necessary to modify the path such that it adjusts to the limits as well as the strengths of the Chinese context and the carrying capacity of the land and resources of China. This situation can be presented as a problem, and a growing problem at that, especially if little of scalable significance is done about it. Though a significant challenge, these conditions represent an enormous opportunity for China – an opportunity unprecedented in history and unparalleled elsewhere in the world. China has the possibility of developing, from the ground up, a new kind of urban-industrial model based on traditional Chinese patterns as well as on China's natural budget. We can assure a durable future only in towns that are based on such a balanced approach.

In this context, we find that it may be appropriate to summarize some of the myriad lessons learned from a multiyear project in rural China that has developed an alternative to China's unsustainable urban development:

- China's unprecedented urbanization program threatens to destroy China's rural villages and their traditional local patterns and culture;
- traditional Chinese villages offer numerous protosustainable models, extensions of which can be used to create sustainable Chinese towns and cities in the future;
- the democratic participatory process called the Sustainable Town Game provides an avenue for synthesizing the best of both old and new, the

traditional and the modern, into a process of democratic planning that can lead to future Chinese sustainable towns and cities;

- the Chinese path to urban sustainability, if it comes, will not follow the model created by many European towns and cities; and
- alternative scenario-building processes using off-the-shelf or on-the-horizon software have a key place in guiding Chinese rural villages to a meaningful and balanced alternative (sustainable city-regions) to China's current megacity-building projects.

In any period of rapid expansion, humankind takes the risk of creating an economic "bubble" that has no other possibility but to burst. Fortunately, resources and the technology now exist to develop an alternative. China desperately needs a major, visible demonstration of sustainability; any of the seven villages studied in the "SUCCESS" project or villages like them could be ideal sites for such a demonstration using principles, processes and new urban models that would demonstrate sustainability for the rest of China and the world. These new sustainable town-regions, as part of ongoing Chinese urban experimentation, will demonstrate a better way and, ultimately, the only way that the Chinese economy can expand to create *zhao kang* (a decent quality of life) in the years ahead.

Chapter 12

THE LONG MARCH TO SUSTAINABILITY IN CHINA

Introduction

China, the sleeping giant, slumbers no more. Whether in climate change strategies, the global food crisis, economic globalization, growing world ecological scarcities (including oil) or the urban-rural balance, China is a major player in exacerbating, mitigating or overcoming these portentous twenty-first century problems. In its drive toward world-class status and its fever pitch efforts toward economic development, this Asian powerhouse has raised its voice in the global dialogue on sustainable development. Even as its city-building programs have moved with unrelenting speed to construct a projected 300 new cities between 2005 and 2010 (Jao, 2005), it has looked both inwards and outwards to experts with ideas for fostering economic and cultural development within its rural countryside.

One such program was the Oikodrom – the Vienna Institute for Urban Sustainability's Sustainable User Concepts for China Engaging Scientific Scenarios (SUCCESS), highlighted in the previous chapter. SUCCESS's research and policy strategy was in many respects convergent with the growing thrust of various Chinese projects and programs initiated across its landmass. Over the last ten years or so, these programs have come to view the village as a focus of rural development work in China and as the appropriate scale for intervention in China by native political elites and domestic and foreign scholars (Brugmann, 1996; Hu, 2008; Stone, 2006). In particular, in the context of mounting natural resource pressures and rising inequalities between Chinese cities and rural villages with respect to standard of living and quality of life, such policymakers have increasingly embraced the idea that the Chinese ecological village serves as an alternative model to prevailing forms of rural development in China (see Hu, 2008, in particular).

Amidst the growing popularity of the village scale – specifically the eco-village – as the model for rural development in China, this chapter offers a case study of a showcase eco-village plan quite dissimilar in design and planning process from those featured in the SUCCESS project. It focuses upon Huangbaiyu, a small village in the east Liaoning province of China, and presents a critical analysis of this village's aspirations and hopes and the designs and implementation program that sought to realize them. It also points to its ultimate failure and serves as a lesson of how a particular understanding of sustainability – developed by William McDonough and Michael Braungart and promoted as a collaborative partnership among its foreign designers (McDonough and Braungart), the Chinese national government (symbolized by the support and oversight by Madame Deng Nan, daughter of Deng Xiaoping), an NGO (the China-US Center for Sustainable Development), a local developer (Dai Xaiolong), a major university (Tongji University) and a local planning group (Benxi Urban Planning Design Association) – was introduced at the local level and played out according to the local-national-global dynamics of power and self-interest among its key players. In this chapter, our major objective will be to answer three questions:

(1) What were the key elements of the McDonough and Braungart "cradle-to-cradle" (C2C) design approach to town sustainability fashioned for Huangbiayu?
(2) What were the sources of the untimely and abject failure of its implementation?
(3) What does Huangbaiyu's fate say about the state of sustainability planning in China?

Huangbaiyu: From Theory to Design

Background

In its ongoing economic development campaign to catch up to and overtake Western capitalist nations and become modernized, Chinese political elites and planners have instituted a string of five-year plans. Given the problems and controversies over accurate data-reporting growing out of the previous five-year plan, China launched its eleventh five-year plan, intending to set the direction for national development from 2006–10, with great hope and ceremony (Fan, 2006; Hum, 2006; Naughton, 2005; and Ziegler, 2006). Among its key themes were several bearing specifically upon rural village sustainability. One such refrain was the idea of the "new socialist countryside" (Fan, 2006) Previous five-year plans had contributed to a growing, even yawning, gap between city and countryside that had produced attendant tensions rising from those inequalities of income, job and educational opportunities, standard of living and overall quality of life. In pursing the new socialist countryside, this new economic program promised to close these inequities and improve the lives of the rural peasantry. Another supportive theme was the effort toward a "balanced development" (Hum, 2006) between cities and rural villages, a program that had been woefully neglected in China's unrelenting campaign to hasten its pace of national economic development during much of the second half of the twentieth century. A third theme – this one bearing on sustainability and social relations – was the concept of the "harmonious society" and its equation in the plan with the idea of a "resource-saving, environment-friendly society." Yet a fourth theme in the eleventh five-year plan was that of the "cycling economy" grounded in the five R's: "rethink, reduce, reuse, recycle and repair" (Naughton, 2005). Finally, underpinning all of the previous ecological and social themes was the primacy – indeed, the continuing hegemony – of the "scientific developmentalist worldview" (Ziegler, 2006). Despite the communist ideology's ambivalent and sometimes tense relationship with modern science, the Chinese communist political leadership remained committed to science as a basic guide and to technology as indispensable tool for moving this former feudal order forward into the new century and toward the goal of economic modernization and its own evolving brand socialism.

In the unfolding of efforts to transform Huangbaiyu into a model eco-village for emulation in other small rural communities across China, these elements of China's latest economic master plan came into play, even as these prominent refrains concealed as much as they revealed about their specific meaning and concrete implementation.

The origins and purposes of Huangbaiyu as a paradigmatic eco-village emerged out of the coincidence of the new strategic emphases by China to respond to mounting evidence of impending natural resource scarcities and to balance development of its rapidly expanding cities and its hitherto neglected countryside; the corporate desire by major American and multinational firms to gain entrée into an immense, emerging Chinese market; the eco-architectural quest for one or more demonstration sites for testing McDonough and Braungart's "cradle-to-cradle" approach to sustainable design; and efforts to blend local, natural, human and financial capital into a model eco-village program.

Even as national political planning elites in China have embarked upon an unprecedented movement of citizens from rural villages to new cities being constructed to meet population growth demographics and shift more Chinese inhabitants from villages to cities, a countertrend has also been in evidence. China's recent focus has been on urban-rural balance and waves of programs dedicated to improving the environment and quality of life of village dwellers through civilized ecological villages such Rizhu, Hainan, Huangbiayu and

Dongtan (Hu, 2008; Williams, 2006; and Ziegler, 2006). This latter policy orientation has been characterized by top-down planning; the search for replicable models or templates capable of being emulated in other villages across China; a dependency upon the use of local resources (venture capital, labor and natural resources) or national ones; and the pursuit of Western models of village planning and foreign assistance (knowledge, expertise and financing). Such Chinese political interest has also been shaped by a number of longstanding and sometimes fairly new cultural-ideological biases favoring ecological modernity, ecological engineering, cleanliness, ideas of progress and "civilization" and an emphasis on material infrastructure (roads, buildings, water and energy systems) over human capital and training and education (ibid.).

A creature of cross-national necessity, the CUSCSD emerged in 1999 as an international public-private partnership dedicated to linking academic institutes and faculty teams, governmental and nongovernmental and industrial and corporate leaders from China and the United States in the common endeavor of seeing China develop along a sustainable pathway. Through the strong directing hand of noted architect William McDonough, the center turned its attention to forging an agreement among the China Association for Science and Technology (represented by Madame Deng Nan) and the CUSCSD (represented by William McDonough). In consummating this deal, the center brokered the Chinese interest in gaining expert knowledge, Western "green" technology and hoped-for financial backing and the primary business interests of the center's corporate sponsors and benefactors in obtaining entrée into the vast and potentially enormously lucrative Chinese market for their products and services. For all parties, interests in mobilizing favorable international publicity, showcasing new technological and design approaches and "successes" and developing a template or model from this demonstration village for replicating its expected success to other poor villages seeking rural development over the rest of the country remained paramount.

Theory and design

The theoretical inspiration for the design features of the Huangbaiyu master plan can be found in the C2C approach formulated by architect William McDonough in collaboration with chemist Michael Braungart. As outlined in their book (McDonough and Braungart, 2002b) and numerous journal articles and internet sites (McDonough and Braungart, 1998, 2003; McDonough, 2007; McDonough Braungart Design Chemistry (MBDC) , n.d.), the C2C's new vista for the next industrial revolution in design and production is fully guided by ecological principles and premises. As disclosed in Table 12.1, McDonough and Braungart seek to carry the governing principles and assumptions of natural ecology beyond the

Table 12.1. Three production design strategies compared

Retroactive production design	Eco-efficient production design	Eco-effective production design
Puts billions of pounds of toxic materials into the air, water, and soil every year	Releases fewer pounds of toxic material into the air, water, and soil every year	Introduces no hazardous materials into the air, water, or soil
Measures prosperity by activity, not legacy	Measures prosperity by less activity	Measures prosperity by how much natural capital we can accrue in productive ways
Requires thousands of complex regulations to keep people and natural systems from being poisoned too quickly	Meets or exceeds the stipulations of complex regulations that exist to keep people and natural systems from being poisoned too quickly	Does not require regulations designed to stop us from killing ourselves too quickly
Produces materials so dangerous that they will require constant vigilance from future generations	Produces fewer dangerous materials that will require constant vigilance from future generations	Produces nothing that will require future generations to maintain vigilance
Results in gigantic amounts of waste	Results in smaller amounts of waste	Results in no waste
Puts valuable materials in holes all over the planet, where they can never be retrieved	Puts fewer valuable materials in holes all over the planet, where they can never be retrieved	Puts no valuable materials in holes anywhere on the planet, since natural and technical nutrients are continually being recycled or "upcycled"
Erodes the diversity of biological species and cultural practices	Standardizes and homogenizes biological species and cultural practices	Celebrates the abundance of biological and cultural diversity and solar income

Source: adapted from McDonough and Braungart, 2002b, and Hawken, Lovins and Lovins, 2000 now-antiquated and destructive features of the first industrial revolution, which was spearheaded by capitalism and superseded even the underlying elements of eco-efficient design and production advanced by Amory Lovins and Paul Hawken (Hawken, 1994; Hawken, Lovins and Lovins, 2000) in their natural capitalism manifesto and tomes. Whereas the eco-Calvinist approach of Lovins and Hawken[1] seeks to squeeze out greater and greater efficiencies in industrial production and energy and other natural resource use (leaving less and less waste and fewer toxic byproducts in the air, water and soil), the eco-effective approach of McDonough and Braungart strives to generate a design and production process that releases no waste or toxic materials into the larger environment, thus mimicking biological or organic systems. Borrowing from industrial ecology, these innovators premise their eco-effective strategy of design and production on two types of metabolisms and two types of nutrients. Biological metabolism refers to the process where complex ecosystems absorb biological nutrients produced by natural processes and whose byproducts provide "food" for other elements or participants of those ecosystems (McDonough and Braungart, 1998 and 2002a). Nature generally does a good job of recirculation. Many technical products of the industrial process however cannot be absorbed and recirculated through these complex, interdependent ecosystems without damage or injury to its functioning. Therefore, McDonough and Braungart argue, industrial design and production processes must be fashioned so that that products and byproducts of systems "composed of materials that do not biodegrade should be designed as technical nutrients that continually circulate within closed-loop cycles" (McDonough and Braungart, 1998: 3). Thus, biological nutrients will be designed to go back to – and be absorbed by – the organic cycle. Technical nutrients, instead of being routinely discarded, will be fashioned so they reenter the technical cycle for continued use as a resource of redesign and production. These eco-effective advocates call the latter process "upcycling" (ibid.).

In the case of the Huangbaiyu master plan (McDonough, 2007; McDonough and Braungart, 2003: 16; and CUSCSD, 2006), McDonough sought to adapt the product design scheme (outlined above) to his town and building plan for this traditional Chinese village of some 1,529 inhabitants. Building on the core idea of the C2C model (i.e., that "waste equals food" (McDonough, 2007: 6)), McDonough's plan called for the industrial byproducts and the agricultural remnants to be captured and reused as technical and biological nutrients in two closed-loop cycles, while carefully segregating them to prevent cross contamination. In the original plan for the demonstration village devised

1 For an overview and commentary on Hawken and Lovins's eco-Calvinism, see Yanarella and Levine, 2008.

by McDonough, local building materials (wood, straw bales, compressed earth) were selected to conform to the biological nutrient requirement of the C2C approach. A biogasifier linked to each villager's house was intended to provide area heating and cooking sources using human and animal waste as the biofuel.

The master plan also incorporated a number of design elements. In order to increase agricultural land, the town's blueprint called for increased village population density, which was aimed at drawing villagers away from the scattered housing of the old Huangbaiyu and concentrating them in residences built within close-knit rows separated by fairly wide streets. This scheme would also permit former dwelling sites to revert to agricultural production sites. Each house had a small garden area for growing vegetables or flowers. Solar panels, donated by one of the corporate sponsors, were placed on a single demonstration house, and the houses were positioned to face south to take maximum advantage of the sunlight. The plan also called for water to be piped to each domicile from a community well. Initially, 42 eco-houses were to be built in the first phase, with a total of 400 eco-dwellings projected to accommodate the 1,529 villagers (Stone, 2006; Sudjic, 2006; and Wang, 2006).

A "SUCCESS"-ful Critique of Huangbaiyu's Failure

With great local, regional and national fanfare, much international attention, and boosterism by promoters and the mass media, Huangbaiyu's demonstration eco-village plan was rolled out and invested with enormous promise and expectation. The outcome of this "green" project, however, proved to be abysmal. The summary remarks on the C2C-driven initiative by Shannon May, the anthropology graduate hired by Intel to study and monitor its development, were chastening:

> A program meant to improve the quality of life of 1,529 villagers by optimizing land use while demonstrating viable forms of alternative energy and green building practices [has been promoted as a success] when there is no alternative energy, the buildings are neither green nor affordable, and empty buildings sit on what was once prime farmland. (May, 2007: 60)

May's somber evaluation raises the basic question of the source(s) of failure of this much-ballyhooed C2C initiative. Did its deeper problems lie in its implementation by local and regional elements of the sustainable development team? Was the national Chinese regime's strategy of top-down planning and bottom-up capitalist development in the rural hinterland a key factor in its demise? Was there a fundamental flaw in McDonough's master plan and/or its underlying eco-effective framework? Was it some of the above, all of the above or something quite extrinsic to these shaping forces?

Implementation flaws

The position of its master architect and the CUSCSD was that the guiding hand of the local developer and building management team was responsible for the implosion of the original eco-village blueprint in its implementation phase (Streeter, 2006; Lesle, 2007; and Toy, 2006a and b). One serious difference that developed between the town designer and the villagers was whether new local industry would be spawned in new Huangbaiyu and who would finance such economic activity. Working through a prominent businessperson and local developer, Dai Xialong, McDonough presumed that Mr Dai would attract capital from other wealthy Chinese to fund local businesses in this showcase eco-village. Many of the villagers, however, presumed from the many publicity tours of the developing demonstration village that foreign corporate representatives and high-level national political leaders and economic planners were prepared to invest in such enterprises that would generate needed jobs to pay for the expensive houses (Buderer, 2006a and b; Toy, 2006a and b; and May, 2007). Nevertheless, this bone of contention did not constitute the heart of the deficiencies in the plan and its implementation.

It is clear from the discrepancies between the original plan and its site layout and buildings constructed in phase one of the project that Dai Xialong took major liberties. Indeed, these departures were not minor. The entire tract of village houses did not face south to capture benefits of active solar technologies that might later be fitted to all homes (beyond the demonstration home). All of the first phase houses included a garage even though no villagers owned or could reasonably be expected to purchase an automobile. Although pipes were laid, the village biogasifier was not completed, purportedly due to the developer's exhaustion of local capital. The yards, which were intended to provide space for individual gardens, proved too small to grow enough food for the families. The yards also proved to be too confining to provide space for the villagers' goats and pigs, which had been raised in the old and scattered Huangbaiyu cottages and used as sources of food for family consumption or sold for income. The housing redesigns and add-ons increased the market price of each so-called eco-dwelling to 1,000,000 yuan – twice the cost of an ordinary village house (Buderer, 2006a and b; May, 2007; Shafer, 2007; Schifferes, 2007; and Streeter, 2006a and b). Only 3 of 42 phase one houses were constructed with C2C required materials (locally grown hay and pressed earth); the rest were predominantly made of hay and compressed coal dust, thus constituting a potential environmental hazard to future village homeowners. Meanwhile, anticipating the proliferation of Huangbaiyu's hybrid design to other Chinese villages, Mr Dai registered trademarks for supposed eco-friendly construction materials – Huangbaiyu roofs, windows, floorings, doors, even Huangbaiyu coal, cameras and cars. Sadly, this entrepreneurial zeal did not match up with successful recruitment of local business development interests and venture capital sought by villagers to enlarge their local economy and bring in badly needed jobs outside of the agricultural sector (Toy, 2006a and b).

Flaws in the eco-effective design strategy itself

On its face, the McDonough–Braungart eco-effective design framework offers a significant advancement over the retroactive production-design approach of industrial capitalism, and even a step or two beyond the eco-efficient production-design orientation championed by Lovins and Hawken. Its virtues also converge with some of the notable elements underpinning the sustainability framework adopted in the five-year SUCCESS program that was implemented in seven Chinese villages. For one thing, the rudimentary developments in village protosustainability fostered by SUCCESS were anchored in the model of natural sustainability and in important respects were equivalent to or isomorphic with McDonough and Braungart's underlying biomimicry model. To wit, natural sustainability has been conceptualized as

> a condition where complex ecosystems such as rainforests, mature water systems, and traditional agricultural systems achieve homeostatic balance or what ecologists call 'climax systems' of high diversity, large biomass, and high stability through protection from rapid change and through shifts of energy flows away from production and toward the maintenance of the system itself. (Yanarella and Levine, 1992c: 309)

Similarly, McDonough and Braungart speak of their underlying foundation as growing out of the idea that "waste equals food" in terms of biomimcry. Biomimcry has been defined in terms of "biomimetic systems [that] are [seen as] closed-loop lifecycles, where outputs and by-products become inputs for something else" (Benyus, 2002: 92).

In addition, both SUCCESS and the eco-effective framework agree on the scope of the problem in China – i.e., its pressing need to house 350–400 million Chinese citizens over the next decade – in a century of mounting global ecological threats and amid rising expectations among developing nations (especially the rural populace). In this respect, both research programs share a belief in the value of starting at the rural-village scale as a counterpoise to the dominant trend, which has favored building large new cities over the past decade in China's urban policy. SUCCESS sees such village developments as protosustainability projects that will eventually need to rise to the scale of the city-region to become sustainable (Levine et al., 2007). McDonough's C2C habitat design, epitomized by Huangbaiyu, seems to be focused on designing an eco-village model that can be replicated elsewhere in rural China (McDonough, 2007).

These areas of convergence should not deflect our attention from the more glaring shortcomings of the C2C production-design approach itself.

At least since 1992, McDonough and Braungart have been working to refine the C2C anchor points into a viable design strategy for cities and towns. In the Hannover Principles they presented that year, McDonough and Braungart outlined a set of first principles to guide the city of Hannover, Germany, in its preparations for hosting the 2000 World Expo (McDonough + Partners, 1992). Partly because these elements did not congeal into an explicit urban design strategy and partly because their C2C framework had not yet crystallized in 2002, they chose a consulting contract with the City of Chicago as an opportunity to grapple with elevating the focus of eco-effective design from the scale of consumer products (e.g., carpeting) and buildings to the scale of a city – in this case, the Chicago metropolis (McDonough, 2002a and b; McDonough and Braungart, 2002a). The title of their work was "The Chicago Principles: Design for Eco-effective Cities." This new and more articulate urban iteration was scheduled for delivery to their clients in 2003, but has never been released (McDonough, 2003, 2004).

The failure to deliver this document and the supporting writings, which supposedly limned these C2C-inspired urban design principles, suggests that a design for effective cities and a strategy for its implementation remain elusive and more formidable than anticipated by the consulting firm McDonough + Partners – Architecture and Community Design. Assuming that McDonough and Braungart's notion of eco-effective cities is their equivalent to sustainable cities or city-regions, their product design principles apparently do not translate easily into the design framework for eco-effective cities.[2]

Part of the problem, we argue, is that McDonough and Braungart have thus far failed to engage the issue of scale in relationship to a theory or strategy of sustainability. Still holding to the Brundtland Commission's minimalist definition of sustainability – as evidenced in the publication of the Hannover Principles – they revert to the less-precise term "green" to finesse the problems of definition and scale. That is, these advocates of eco-effective design simply assume that such a design framework can, and should, be applied at every

2 Indeed, the only product of this design intervention has been green roofs designed by McDonough's firm in various locations around Chicago, including Chicago's City Hall (McDonough + Partners, 2001).

scale or level without qualification or adjustment. By contrast, SUCCESS privileges the scale of sustainable city-regions, and treats lower scales as merely sustainability-driven or protosustainable and so-called higher scales as extra- or transsustainable. In addition, by grounding design issues upon the eco-effective production-design model, McDonough and Braungart remove or marginalize issues of political participation and stakeholders' negotiations for sustainable urban design from every level of the city – from the neighborhood to the metropolis. In so doing, they reveal the penchant among many architects and planners to adopt a technocratic model of governance that often is augmented by a cozy relationship with corporate power.

These general urban design strategies, we argue, resulted in negative consequences due to the failed effort to convert eco-effective design imperatives into a strategy for the (proto) sustainable eco-village construction in Huangbaiyu. The first place where the McDonough village design approach went wrong was in its tendency to elevate design imperatives and market values over and above participatory planning imperatives and democratic values. That is, even in the absence of the deficiencies emerging from the gap between eco-architectural designer master plan and developer implementation, the plan itself did not incorporate any significant and meaningful role for the villagers' participation in the design and implementation processes. No one ever held any village-wide meetings of Huangbaiyu residents or actively engaged these local stakeholders in proposing, evaluating or contributing to the design and building of the project. No one ever established a mechanism of accountability to villagers so that the village community could assess and influence any gulf between village design and construction (May, 2007; Lesle, 2007). Symbolic of this cavalier attitude toward community participation, no one incorporated a public square or common meeting place into the master plan.

No less serious, no one paid deference in the overall process to local knowledge of village topography, social relations and the village-agricultural countryside interplay and no effort was made to incorporate the villagers' sustainability practices in farming and fishing into the master plan or village design (Lesle, 2007). Instead, villagers were treated as passive subjects to be merely informed or manipulated for their own "good." There was no buy-in from the villagers to the process; therefore, villagers exhibited no sense of ownership toward the eventual outcome of the project. By contrast, SUCCESS recognized the instrumental and intrinsic rewards of active villager participation in each of the seven sites involved in its action-oriented research. In general, such village involvement was seen as an integral part of the Sustainable City Game (Yanarella and Levine, 2000) constituting the process of uncovering each village's community needs and building multiple scenarios for responding to them.

Table 12.2. Typology of community participation in development projects

Types of participation	Features of each type
Manipulation	Participation is a pretence, local people have no power
Passive	People are involved by being notified of what has already been decided; unilateral announcements are made by project managers without listening to peoples' responses.
Consultative	People participate by being consulted or by answering questions. External agents control analysis and decision making
Participation for material incentives	People participate by contributing resources such as labor in return for food, cash or other material incentives; people have little stake in prolonging practices when the incentives end
Functional	Participation is seen by project managers as a means to achieve project goals, especially reduced costs. This may involve shared decision making but tends to arise after major decisions have already been made by external agents. At worst, local people may still only be coopted to serve external goals
Interactive	Participation is seen as a right, not just a means to achieving project goals; people participate in joint analysis, decision making, the formation of action plans and the formation or strengthening of local institutions
Self-mobilization	People participate by taking initiatives independently of external institutions to change systems. Contacts with external agencies are made for access to resources and/or technical advice needed, but control over how resources are used is retained by the group. Such self-initiated mobilization may or may not challenge existing distributions of wealth and power.

Source: Pretty, 1995: 1252

Based on Pretty's (1995) community participation typology (see Table 12.2), the activities of the SUCCESS Chinese village project routinely reached Level 6 ("interactive") forms of citizen participation and strove for Level 7 ("selfmobilization") manifestations.

Onsite observations and reports by May and others bear out the conclusion that C2C villager participation never rose above Level 3 ("consultative") practices on the Pretty typology. Framed largely within market values and top-down/elite decision-making orientations, McDonough's program lacked an analogue to SUCCESS's version of the Sustainable City Game and evidently saw no need for one. Neither did the C2C master planning and implementation process see any virtue in capacity building and efforts to generate social capital among the villagers. As a result, villager consultation was simply a pro forma exercise with no practical purpose beyond (perhaps) marshaling citizen legitimacy toward the elite-driven activities and decisions. Only minimal general information about the project and its intended community benefits was conveyed to villagers in an attempt to garner public credibility and passive support. Despite hollow claims to the contrary (see photos of village "meeting" in Wang, 2006), the Huangbaiyu consultative process merely substituted a weak representative model of civic participation formed by its involvement of Dai Xaiolong as the putative village leader and by the chief financial backer as head of the implementation team. In this respect, the C2C market and governance orientation toward the project paralleled the Chinese national elite penchant for top-down planning; the search for replicable models for widespread application; dependence on local financial and other resources; and the "wager on the strong and successful" over the weak and poor (Hu, 2008; Martens, 2006; and esp. Ziegler, 2006).

Another shortcoming integral to the C2C design model and community strategy relates to its failure to come to grips with the meaning of sustainability in general, in particular at the village, agricultural land scale. SUCCESS was grounded in the idea that sustainability can best be achieved as a balance-seeking process at the scale of the city-region. This claim is premised on notion that the city-region is

> the largest unit capable of initially addressing the many urban architectural, social, economic, political, natural resource, and environmental imbalances besetting the modern world and, at the same time, the smallest scale at which such problems can be meaningfully resolved in an integrated and holistic fashion. (Yanarella and Levine, 1992b: 771)

Moreover, given its commitment to strong sustainability and its distinction between economic growth (which is quantitative in nature) and sustainable development (which is qualitative in character), SUCCESS is fundamentally concerned with working within the ecosystem's carrying capacity and striving to advance urban sustainability as a qualitative process of true development.

The social and economic implications that McDonough and Braungart draw from their eco-effective production-design model run counter to SUCCESS's position. In particular, C2C's underlying market ideology reopens the growth/ development debate within the sustainability literature by unapologetically embracing growth and consumerism. For an eco-architect, McDonough took the intellectually audacious position in a recent interview that "the growth/no-growth argument is specious. Growth is good. The question is, how do you want to grow?" (Surowiecki, 2002: 1). A no less contrarian claim was attributed to McDonough when he said we don't need to make less stuff. We only need to make stuff different (Surowiecki, 2002, evidently paraphrasing McDonough). While one might excuse this as a bit of popular rhetoric designed to win over resistant business representatives and consumers to sustainable living, it is fair to conclude that McDonough's entrepreneurial gusto prompted him to see his C2C design as an approach to a new way of reinstituting industrial production. His method averts ecological scarcities and ecosystem limits and promises the renewal of the American Dream and its extension to the inhabitants of developing nations. Researchers may wish to examine to what degree the Huangbaiyu-C2C strategy blinked at major issues of class, gender and ethnicity in imagining the trajectory of the presumed social development of this eco-village and inhabitants in twenty-first century China.

McDonough's (2007) C2C-informed master plan of the new Huangbaiyu is steeped in many of the assumptions and operating premises of ecological modernization. In the context of the Huangbaiyu eco-village initiative, both McDonough's approach and some versions of ecological modernization share a number of key features. These include an unremitting commitment to modernity; a strong devotion to the role of technology in solving social and environmental problems; and a strong belief and overt support for the capacity of corporate and transnational capitalism to overcome structural deficits that have hitherto devalued natural capital and exploited nonrenewable natural resources. This approach seems anchored in the principle of "sustainable growth," which is founded upon the unlimited human consumption of natural and human capital at enormous social and ecological costs.

McDonough's and the CUSCSD's complicity in national and global corporate vehicles of unlimited capitalist growth and global-market penetration has already been alluded to in the deeper motives of the sponsors and benefactors of McDonough's Chinese endeavors and the CUSCSD. The desire to achieve access to high-level Chinese political and economic planning elites and thus to gain a foothold in China through the Huangbaiyu project seems apparent.[3] Despite the Huangbaiyu fiasco, the CUSCSD has accumulated $50,000 from each of the corporate members of its "Founders' Circle," a list of almost two dozen corporations that includes Dow, Ford,

3 Shannon May's incisive, ground-level critique of McDonough's project is deepened in her more recent articles – especially May 2008a and 2008b.

Vermeer, Intel, Hewlett-Packard, BASF, BP and Autodesk (May, 2007; and CUSCSD webpage).

Conclusion

It is hard to derive any other conclusion from this initiative than that Huangbaiyu has been an unmitigated failure. Chinese and foreign sponsorship tours have dwindled to zero. Local, national and foreign financial investment has completely dried up. The old, decentralized Huangbaiyu has remained the social space of village interaction, traditional farming and local-regional commerce, while the new Huangbaiyu has become little more than a cemetery to dreams that will never be realized. The "sustainable" village remains unpopulated, except for two families occupying two of the houses. The project's internet presence, so much a magnet for promoters and other supporters of its possibilities, has been effaced from CUSCSD and McDonough web pages as if it had never existed.

Meanwhile, William McDonough + Partners and CUSCSD have taken leave of Huangbaiyu and proceeded to design innovative projects for sustainability in Shanghai and Nanjing, and for new urban townships to absorb the 350–400 million population increase in Chinese cities and towns over the next 25 years. Shannon May, a University of California at Berkeley anthropology graduate student, has moved on to finish her dissertation and write a book on the lessons of the of the Huangbaiyu project for future rural and sustainable development. Chinese elite interest has shifted its energies and efforts to Dongtan, near Shanghai, where the global design and project management firm ARUP has assembled a "green" master plan for converting Dongtan into another showcase urban sustainability project – another possible national prototype – and a new opportunity for unparalleled ecological achievement or monumental failure (McGray, 2007; McKenzie, 2007; and Girardet, 2006).[4] Phase one, expected to be completed by 2010, was projected to bring an influx of 10,000–25,000 people into an evolving "sustainable" city which will concentrate a vast array of "green" architecture, alternative energy and conservation techniques and technologies and fuel-celled vehicles, etc. and will eventually grow to 500,000 residents (Normile, 2008; Ruiz, 2006). Sadly, ARUP's efforts have stalled and the new showcase of Chinese urban sustainability has fallen on hard times (Huang, 2009). While it is perhaps too

4 Note the critical comments on Dongtan by ecological footprint pioneer, William Rees, who sees this proposed new Chinese sustainability prototype as "hardly a truly sustainable option," given its construction on former agricultural land and an ecologically significant wetland (Ruiz, 2006).

late to influence the development of Dongtan, it may be profitable to close this case study by enumerating a set of critical lessons derived from Huangbaiyu debacle.

The intrinsic and instrumental value of strong civic participation across entire planning and construction process. Huangbaiyu illustrated what happens when professional architects, designers, developers and planning officials keep village inhabitants at arm's length in all phases of a rural development project. Important native resources are neglected, and real and perceived needs of citizens are excluded from the larger and official discourse of planning.

The priority of capacity building and social capital over material outputs and infrastructural improvements. Both Chinese national elites and foreign professionals exhibited a preference for material outputs (buildings, houses, utility lines) and infrastructure improvements (roads, landscaping, etc.) over fostering public education and the civic virtue and citizenship tools of its villagers. As a result, they made significant design errors and sidestepped and wasted potential citizen resources. Not least of all, no one advanced or enhanced the tools of democratic governance and community development.

The necessity of strong over weak sustainability. Whether Huangbaiyu could have become a national template for widespread replication of eco-village development is an open question. What is clear, however, is that the translation of C2C into a village design strategy diluted the requirements of strong and robust sustainability and substituted an approach to community design whose form emulated that of a small suburban housing tract.

The synergistic merits of outside expertise and local knowledge. The village families of Huangbaiyu had lived for many decades in their village habitat and rural surroundings. In the course of their inhabitation, they accumulated valuable local knowledge about its local ecosystem and how to appropriate sustenance from the land, streams and rivers without exploitation. That knowledge was as integral as the scientifically derived knowledge generated by specialists and consultants who dominated the design and master plan of Huangbaiyu. A truly sustainable project seeks to productively integrate these different, but inherently useful and prized, resources.

The necessary balance between village and countryside, city and region. As the design of the new Huangbaiyu unfolded, villagers complained of the greatly increased distance between the new village and their agricultural plots. They also criticized the ill-fitting and overly small size of the garden area, either to raise adequate vegetables for their families or to raise their animals for slaughter and meat. Obviously, villager participation in the design and planning process would have obviated the analytic separation of village and farmland made by Western experts in shaping Huangbaiyu's master plan.

The primacy of the city-region as the telos of sustainability-oriented village development. Because McDonough and his professional design assistants worked from too small a scale – the scale of the village – and because they embraced a conception of sustainability that remained abstract and elusive in some ways, their design and implementation strategies operated at the sustainability-oriented level at best. Consequently, the transit from a protosustainable village to a genuine sustainable city-region, which was an integral facet of the SUCCESS research program, was precluded from the outset.

One additional point is worth noting. The SUCCESS project was a research project aimed at exploring the characteristics of protosustainability and projecting them as sustainability into the future. As such, no one expected that an actual sustainable village or town would be an immediate outcome. The Huangbaiyu project, on the other hand, bore a strong resemblance to a normal professional building design assignment (although its muse may have been a somewhat elusive and undefined quest for sustainability). Such resemblances die hard. From our current perspective, it is difficult to assess which of the many actors and influences bore the preponderant responsibility for the failure of the project. There were, to be sure, numerous stakeholders and decision makers – many of them still unmentioned or unknown. From a prominent ecological architect to an international organization, from the local developer to the international sponsors and from the village governance process to the many higher levels of Chinese government and agencies with jurisdiction over the project, from material suppliers and actual builders, and so on – all of these actors had at least somewhat ambiguous and overlapping roles on this new sort of project. By contrast, the strength of SUCCESS was in its insistence on an extensive, ongoing stakeholder participation process, with numerous feedback loops to engage all who could be engaged, and also in its resolve to elicit buy-in to the process itself. SUCCESS also operated through a strong, systemic, operational definition of sustainability where ideas and values were continually tested and subjected to balancing at the scale of the village system as a whole. At Huangbaiyu, many simple mistakes and unfortunate decisions were made in the development process without ever apparently being tested or subject to feedback from other stakeholders – until it was too late.

Numerous aspects of these two projects appear to be potentially complementary. One can only wonder what sort of project might have emerged had the strong participatory process operating within the clear, systemic, sustainability definition of SUCCESS could have grounded the underlying framework of such a well-sponsored development project that Huangbaiyou started out to be.

As putative sustainability megaengineering projects in Dongtan, Dubai and Vancouver overshadow the demise of the hopes and dreams of the little village of Huangbaiyu for a better future, the lessons of this social experiment and urban architectural fiasco should not be lost. Looming ecological threats and fateful economic and agricultural tendencies are especially concentrating their impacts upon the shorelines and urban centers of the world. The local-global discourse of sustainability remains a contested one, even by those who speak so passionately on its behalf. In its hopes and in its failure, Huangbaiyu still participates in this conversation, if we but listen.

CLOSING THOUGHTS

Conclusion

TAKING THE ROAD LESS TRAVELED

> I shall be telling this with a sigh
> Somewhere ages and ages hence:
> Two roads diverged in a wood, and I–
> I took the one less traveled by,
> And that has made all the difference.
>
> —Robert Frost, "The Road Not Taken"

Against the background of the imposing dangers to international stability, global ecological equilibrium and human survival, we close this book with a summary of the road *most* traveled to date in the projects and programs to build sustainable cities and city-regions. Despite the welter of "sustainable cities" populating the built environment of the twenty-first century, we argue that only the road less traveled – i.e., the strong sustainability strategy outlined in this book and used as a critical evaluative framework for analyzing representative cases around the world – holds out hope of creating new sustainable city-regions across the globe. Such cities can, and must be, guided by their technological tools, architectural blueprints and sociocultural practices defining strong and robust sustainability. Only this approach offers genuine hope of closing the gap and overcoming the frictions between the nations of the north and those of the south. That is, we can only accomplish this goal by greening the cities of the north and providing exemplary strategies, tools and programs for the developing nations of the south to pursue an alternative and ecologically resilient path to strong urban sustainability. Only then will incrementalist, reformist, smart strategies dominating the sustainability debate be surpassed by the achievement of true sustainability locally that radiates outward regionally, nationally and globally. Only then can many of the deeper causes of international war, conflict and unrest be mitigated or overcome. Only then will the true and historic city become the fulcrum of global sustainability.

Worlds in Collision

The many forces shaping the contours of the twenty-first century and the world's complex and contradictory global landscape have conspired to make the urban sustainability agenda both more difficult and more essential. Global warming has posed the most extraordinary challenge to international organizations, nations, towns and cities, not only because its mounting impacts are palpable, but because the strategies for overcoming it confront the continued "imperatives" of growth in the north and the south from transnational corporations, which still operate under shareholder demands for steady and substantial dividends. Corporate CEOs and boards seem fixated on market share and profits (McKibben, 2010). The standard response from these corporations, with few notable exceptions (Hawken, 1994), are pseudosustainability campaigns that involve varieties of "greenwashing" that at best temper protests from outsiders and renegade stockholders. However, these campaigns dramatically fail to restructure overall goals and motivations of global business.

Compounding the ascending hazards of greenhouse gas buildups is the imminent problem of peak oil (Kunstler, 2006). Advertising promotions by the oil majors paint a positive image of supply-side solutions to an incipient problem of epochal proportions by pointing to the latest magic bullet of deep oil drilling technology and the complementary role of natural gas exploration and discovery deep underground or beneath the oceans. Yet they fail to acknowledge that even the discovery of an uncertain number of Prudhoe Bay equivalents will not satisfy increasing global energy demands by China, India and other developing countries or overcome sagging world oil reserves caused by the phenomenon of peak oil and the downward slide that follows it. Both oil and natural gas are finite and therefore depletable energy resources. Even if M. King Hubbert and other knowledgeable petroleum experts are slightly off in their estimates, the groundwork for a postpetroleum world is being laid and will not yield to midterm technological remedies (Deffeyes, 2008). Trying to solve the peak oil problem by way of demand-side nostrums through current technological alternatives – like hybrid cars and more efficient use of oil as an energy source – will merely stretch existing supplies by a few decades. Creating a bridge to renewable energy sources is obviously the strategy to take, but current global thinking suggests that any serious strategy combining conservation with renewables is likely to be too little and too late (Balfour and Keenan, 2007). For the peak oil problem, concentration within the oil industry remains an enormous obstacle to overcoming the formidable economic and political barriers to addressing the end of oil in a rational, sensible collective and timely manner.

The momentous shift in the world's population (witnessed in the first decade of the new millennium) poses itself in ambiguous terms. On the one hand, this migration offers a major portion of the world's population the opportunity to experience the urbanity and individual freedom, the sociability and communal possibilities, the culture and excitement and the diversity and tolerance associated with the city by twentieth-century students of urban life like Max Weber and Lewis Mumford. On the other hand, the city's greatest critics, including Emile Durkheim and William Whyte, have emphasized the negative consequences of the transit from the countryside and the city – among them growth of alienation and anomie, loss of communal ties and creation of an artificial and superficial self and obsession with the lures of consumer culture and the compulsive pursuit of fashion. (On these contradictory orientations toward the city, see: Yanarella, 2001: 123–42.) As Murray Bookchin suggested in his minor classic, *The Limits of the City* (1996), the solution is not to polarize the city and countryside without obliterating important differences between the two in order to bring the city to the countryside and the countryside to the city. (See Raymond Williams' classic work on this topic – Williams, 1975.) Socially and politically, this sublation of the two has been difficult to accomplish in real world settings.

While leading demographers forecast the topping out of world population growth to around 9 to 10 billion inhabitants of the globe by the end of this century, few sustainability advocates should be sanguine about those prospects. The carrying capacity of the world's resources will surely be exceeded by such a figure and, in any case, the geopolitical, religious, cultural and ethnic, natural resource and other conflicts and frictions are likely to be exacerbated by an even more crowded world than we are confronting today. Even if enough resources were to exist to adequately satisfy such numbers, the economic inequalities built into the contemporary world of nations are likely to present stark differences in quality of life, standards of living, life chances and mortality rates.

The US-induced global financial meltdown and accompanying economic slowdown triggered in 2008 posit their own unique challenges for economic well-being and the quest for a new local-national-and-global formula for restarting sustained economic growth. Such a task will be especially difficult in a world where pernicious practices cause partial financial collapse, where deep economic recession has to be overcome and where instituting a "new normal" may prove to be a formidable challenge stemming from certain long-term structural barriers to returning to postwar conditions (the "old normal") of the fifties and early sixties. These conditions postulated an end to ideology (Bell, 1960) then later an end to history (Fukuyama, 1992) and now the supposed new dawning of an enduring era with no limits to growth. That projected epoch now looks like a chimera by a particular ideology trumpeted at a particular historical period with a set of particular and unique social conditions (Stiglitz, 2010). Worse still, the assuming forces now shaping local-global and regional-national dynamics suggest that this Panglossian dream is not merely a pipe dream but absolute folly, even as rightwing groups and movements call for its reactivation, institution and triumph.

Whether we can win the battle with terrorism or must languish in a condition of persisting inequalities of wealth, power and status in an increasingly resource starved and unsustainable world, the war on terror exists as a fear-inducing and material feature of the world we live in today. Although terror has been a political fixture of the modern epoch as an instrument of perceived underdogs in societies structured and governed by top dogs (Fanon, 2005; Sorel, 1961), the level of terrorist activity and its deadliness and expanse may be unique to our circumstances. No doubt, each epoch in the modern history of terror requires its own contextual analysis and nuanced response. Ours has been notable by its clash of religious fundamentalisms and a communications framework that tends to muffle and suppress deeper understandings of its origins and intensity – and likely its potential cures. Aspects of terrorism may have profound foundations resident in human psychology and cultural nihilism. Its abatement, however, may yet be found at least in part in the collective building of a new global structure grounded in sustainable institutions and practices (Levine, Yanarella, Radmard and Dumreicher, 2003).

Sustainability Strategy: The Road Less Traveled

In 1976, Amory Lovins published a highly influential essay in *Foreign Affairs* magazine titled, "Energy Policy: The Road Not Taken?" (Lovins, 1976). This celebrated analysis proposed a way out of the dilemmas faced by the United States and other Western industrial nations through what Lovins called the soft technology pathway of renewable energy and conservation augmented by using coal as a bridge fuel to a world of solar-based renewable energy. While knowledge of the threat that greenhouse gases might have posed for this alternative energy strategy was then either nonexistent or in its infancy, Lovins' soft energy path captured the hearts and minds of many environmentalists and solar advocates in the United States. His proposed course also triggered a serious energy debate here and elsewhere about our energy future in between the two oil shocks of that decade.

Our times are experiencing multiple crises on a variety of geographic scales that call for serious reconsideration and reevaluation of foundational values, social and economic institutions, political practices and visions of the good life and the good society. Such an era also calls for a serious rethinking of society's relationship with a natural world that is deeply implicated with the society and its built environment and in some ineliminable sense separate from it. The message of this volume has been that neither the hegemonic route traveled by mainstream capitalists (and even state socialist societies), nor the weak sustainability byways pursued by even the crown jewels of urban sustainability in North America, Europe, Asia or Africa point the way to the route that must be followed. No strategy, however well-intentioned or generously funded, is likely to meet the formidable challenges presented by our times unless it takes up the conceptual tools and strategic framework flowing from the foundations of the strategy of sustainable cities outlined in this work. This statement is less an expression of hubris or arrogance than a declaration that the conditions and risks posed by peak oil, world population trends, global warming, financial upheaval and terrorism demand a radical (in the sense of going to the roots) and thoroughgoing strategic response.

Weak forms of green technologies and other practices and standards dressed up as sustainability witnessed today do not individually or collectively amount to a revolution. As we noted earlier in this volume, Thomas Friedman (2008: 205) characterized the literature on the x-number-of-lessons-to-a-green-future fashion spreading across America and other countries simply and with utter frankness as little more than a "green party," not a serious revolution. In the hot, flat and crowded world that is crystallizing from the vectors of partly convergent, partly oppositional forces, the green banner leading the parade down Main Street is not going to get us to the just and sustainable communities and city-regions that must be the building blocks to global sustainability. Given the portentous threats and challenges we face, who can really think otherwise?

Those of us who have worked off of Wackernagel and Rees's (1998) notion of a "fair Earthshare" flowing from their ecological footprint analysis have striven to find a calculable sustainability metric for towns and cities and to counter the conflation of economic development and economic growth. But, at the ground level where real people struggle to deploy real strategies and tactics to advance effective political action and enact formal policy, laws and regulations to promote strong sustainability, the sobering truth is that all existing urban sustainability projects *are really founded upon some species of growth management.* The line of demarcation between all forms of quantitative economic growth (including smart growth) and qualitative sustainable development drawn by Redclift (1987) and Daly and Cobb (1994) has yet to be crossed. As the recent judicial rebuff of Pleasanton, California's efforts to establish a population cap within its city borders (Egelko, 2010) and as the analogous attempt by Okotoks, Alberta, (see Chapter 9) to maintain a cap on its population (tied to a maximum annual allotment of water from the Sheep River running through it) have shown, judicial interpretation and regional "growth imperatives" have a way of trumping no-growth ambitions. (To see how formidable a political task is involved in moving beyond growth management toward truly strong and vibrant sustainable cities, see: Zovanyi, 2001.) Others must join those who are tilling the fields of urban sustainability for the political bases of strong urban sustainability and qualitative sustainable development in order to defend and popularize *strong sustainability* and to contribute to fashioning a politics that works to realize an *existentially realized strong sustainability* (see the urban sustainability rubric in Chapter 3).

The concepts and theories, measuring rods and techniques in the toolkit provided in our elaborated strategy of strong sustainable city-regions do not have a pervasive following today or substantial financial backing. That strategy, though tested against alternatives, designed for implementation in real cities, towns and villages and put forth in national and international competitions, is not (yet) popular or in fashion. Nor would it be cheap. It does however overcome many of the evident flaws and gaping holes in weak and cosmetic variants from conventional environmentalism and smart growth to species of weak or showcase sustainability. (Recall here the arguments marshaled in Chapter 3.)

Neither is it complete nor is it without gaps or rough edges. The Sustainable Area Budget is a metric in the making with many technical points yet to be clarified or resolved. The Sustainability Engine, while conceptually complete in many ways, still awaits computer science and software engineering to catch up and fully realize its technological telos. Nevertheless, strong sustainability of the type propounded in this work or advanced by others can be completed by those with the uncompromising will and political determination to accept nothing less than what is actually necessary to confront the daunting challenges of "our times" with others and recover the earthly and public commons from those agencies that are poisoning one or the other – or as likely both.

Is this "tract of our times" merely a utopian representation of a condition or set of circumstances that humankind cannot get to from here (one of the cardinal characteristics of utopian thinking)? Will the strategy it proffers crash against the Charybdis of an inevitably violence-prone and nihilistic politics, built upon a Hobbesian image of basic human nature? Or, will it be shattered by the Scylla of the inevitably individualistic and possessivistic assumptive world of a capitalistic ideology and set of local-global institutions ever reshaping themselves, but never being surmounted or tamed by a new sustainability ethic and structure? In this route less traveled, we prefer to ground hope in our proffered theory and strategy of sustainable city-regions, seeing in them the working resources for an *eminently practical utopia*. The human future is a possible horizon whose human grasp relies on the lessons from past protosustainable social and cultural practices and the emerging possibilities and potential of twenty-first century science and technology. It also relies upon the unquenchable human passion for democratic governance, beloved community and social justice as well as the underlying wisdom of the Earth.

In Conclusion: Is the City the Fulcrum of Global Sustainability?

The city-region is not only a practical utopia. In a world fraught with the looming and portentous ecological scarcities, financial disasters, environmental threats and population pressures implicated in the rapidly changing global system of systems, it is a *practical necessity* – a formidable but necessary component of urban-rural redesign of our built environment, which is entrenched in our natural landscape and the monumental task of political-cultural reconstruction. In the midst of the worldwide expansion of scientific knowledge and technological knowhow in the biosciences and cognitive neuroscience as well as artificial intelligence, information technology, digitization and computer microminiaturization, it is tempting to presume that the novelties of twenty-first century science and technology will win out over the deeper truths and age-old verities of the ecosystem being learned anew by the eco-sciences. (On this debate, see Garreau, 2006; and Lanier, 2010.)What the polis and the medieval hilltown tell us, however, is that (in spite of the prodigious advances of contemporary science and technology) the lessons of sustainability lie buried in the historically conditioned insights and circumstances of men and women of bygone ages where authoritarian power was held in check and technological development was put to the human and civic uses of the larger community (Levine, 1987; and Yanarella, 2001). These historical exemplars also adumbrate the social, political, economic and cultural conditions that give meaning and substance to refounding individual freedom within a larger, more encompassing communal social and political form. These conditions also give meaning to personal creativity within a wider cultural matrix linking individual and collective genius in the practices of human beings in the everyday celebrations of human life and settlement. Such practices and celebrations include production and reproduction, play, language, the human spirit and the eternal mysteries that unite us as human beings with the Earth.

Seen in these terms, the various crises and challenges besetting the world of the twenty-first century beg for solutions that surpass scientific discovery and advanced technological instruments. Naturally, science and technology will play their limited parts in overcoming the ecological, energy, population, economic-growth driven and resource crises of "our times." What troubles us are the conventional and antiquated political nostrums still being pursued to achieve a return to the "old normal" (consumer society, social injustice, alienated politics, global stratification, industrial and corporate agriculture, the generational sacrificial contract, the dream colony of the entertainment industry, the permanent war economy, etc.).

In the face of such nostalgic fantasies, the pursuit of the strategy of sustainable city-regions presents itself as a cultural alternative and a political possibility for breaking through these many entwined socioeconomic and political emergencies. This strategy is in no way a mechanistic Archimedean fulcrum to move the world. Nonetheless, it poses itself as an avenue that today may still be less traveled, but which remains a road that in the face of the looming problems of peak oil and climate change points the way to the kind of paradigm change and political transformation whose tactics and strategies are only beginning to be limned (Reid and Taylor, 2010). If this is so, the world of the twenty-second century will be structured, organized and rebuilt in ways that are very different from that of the late twentieth and early twenty-first century. It will be built upon cultural foundations and underlying assumptions borrowing heavily from the wisdom of Earth and the anchoring truths of permaculture, the knowledge of earlier ages and glorious examples from humankind's highpoints in the past, as well as the fragile and still miraculous secrets of life and the human body (Balfour and Keenan, 2007; Holmgren, 2002, 2009; Hopkins, 2008; Berry, 2001).

These concluding speculations are beginning points for another work, one that reframes the role of the sustainable city-region as the fulcrum of global sustainability upon risks and hazards, uncertainties and imponderables, which reside squarely upon politics and power and yet no less so, as this book has sought to do, upon theory, design and metrics. Permit us voyage.

APPENDIX

Appendix A

CHARTER OF EUROPEAN CITIES AND TOWNS TOWARDS SUSTAINABILITY

(As approved by the participants at the European Conference on Sustainable Cities and Towns in Aalborg, Denmark on 27 May 1994)

- Part I: Consensus Declaration: European Cities and Towns Towards Sustainability
- Part II: The European Sustainable Cities and Towns Campaign
- Part III: Engaging in Local Agenda 21 Processes: Local Action Plans Toward Sustainability

Part I: Consensus Declaration: European Cities and Towns Towards Sustainability

I.1 The role of European cities and towns

We, European cities and towns, signatories of this Charter, state that in the course of history, our towns have existed within and outlasted empires, nation states, and regimes and have survived as centres of social life, carriers of our economies, and guardians of culture, heritage and tradition. Along with families and neighbourhoods, towns have been the basic elements of our societies and states. Towns have been the centres of industry, craft, trade, education and government.

We understand that our present urban lifestyle, in particular our patterns of division of labour and functions, land-use, transport, industrial production, agriculture, consumption, and leisure activities, and hence our standard of living, make us essentially responsible for many environmental problems humankind is facing. This is particularly relevant as 80 percent of Europe's population live in urban areas.

We have learnt that present levels of resource consumption in the industrialised countries cannot be achieved by all people currently living, much less by future generations, without destroying the natural capital.

We are convinced that sustainable human life on this globe cannot be achieved without sustainable local communities. Local government is close to where environmental problems are perceived and closest to the citizens and shares responsibility with governments at all levels for the well-being of humankind and nature. Therefore, cities and towns are key players in the process of changing lifestyles, production, consumption and spatial patterns.

I.2 The notion and principles of sustainability

We, cities and towns, understand that the idea of sustainable development helps us to base our standard of living on the carrying capacity of nature. We seek to achieve social justice, sustainable economies, and environmental sustainability. Social justice will necessarily have to be based on economic sustainability and equity, which require environmental sustainability.

Environmental sustainability means maintaining the natural capital. It demands from us that the rate at which we consume renewable material, water and energy resources does not exceed the rate at which the natural systems can replenish them, and that the rate at which we consume nonrenewable resources does not exceed the rate at which sustainable renewable resources are replaced. Environmental sustainability also means that the rate of emitted pollutants does not exceed the capacity of the air, water and soil to absorb and process them.

Furthermore, environmental sustainability entails the maintenance of biodiversity; human health; as well as air, water, and soil qualities at standards sufficient to sustain human life and well-being, as well as animal and plant life, for all time.

I.3 Local strategies towards sustainability

We are convinced that the city or town is both the largest unit capable of initially addressing the many urban architectural, social, economic, political, natural resource and environmental imbalances damaging our modern world and the smallest scale at which problems can be meaningfully resolved in an integrated, holistic and sustainable fashion. As each city is different, we have to find our individual ways towards sustainability. We shall integrate the principles of sustainability in all our policies and make the respective strengths of our cities and towns the basis of locally appropriate strategies.

I.4 Sustainability as a creative, local, balance-seeking process

We, cities and towns, recognise that sustainability is neither a vision nor an unchanging state, but a creative, local, balance-seeking process extending into all areas of local decision making. It provides ongoing feedback in the management of the town or city on which activities are driving the urban ecosystem towards balance and which are driving it away. By building the management of a city around the information collected through such a process, the city is understood to work as an organic whole and the effects of all significant activities are made manifest. Through such a process the city and its citizens may make informed choices. Through a management process rooted in sustainability, decisions may be made which not only represent the interests of current stakeholders, but also of future generations.

I.5 Resolving problems by negotiating outwards

We, cities and towns, recognise that a town or city cannot permit itself to export problems into the larger environment or to the future. Therefore, any problems or imbalances within the city are either brought towards balance at their own level or absorbed by some larger entity at the regional or national level. This is the principle of resolving problems by negotiating outwards. The implementation of this principle will give each city or town great freedom to define the nature of its activities.

I.6 Urban economy towards sustainability

We, cities and towns, understand that the limiting factor for economic development of our cities and towns has become natural capital, such as atmosphere, soil, water and forests. We must therefore invest in this capital. In order of priority this requires:

- investments in conserving the remaining natural capital, such as groundwater stocks, soil, habitats for rare species;
- encouraging the growth of natural capital by reducing our level of current exploitation, such as of non-renewable energy;
- investments to relieve pressure on natural capital stocks by expanding cultivated natural capital, such as parks for inner-city recreation to relieve pressure on natural forests); and
- increasing the end-use efficiency of products, such as energy-efficient buildings, environmentally friendly urban transport.

I.7 Social equity for urban sustainability

We, cities and towns, are aware that the poor are worst affected by environmental problems (such as noise and air pollution from traffic, lack of amenities, unhealthy housing, lack of open space) and are least able to solve them. Inequitable distribution of wealth both causes unsustainable behaviour and makes it harder to change. We intend to integrate people's basic social needs as well as healthcare, employment and housing programmes with environmental protection. We wish to learn from initial experiences of sustainable lifestyles, so that we can work towards improving the quality of citizens' lifestyles rather than simply maximising consumption. We will try to create jobs which contribute to the sustainability of the community and thereby reduce unemployment. When seeking to attract or create jobs we will assess the effects of any business opportunity in terms of sustainability in order to encourage the creation of long-term jobs and long-life products in accordance with the principles of sustainability.

I.8 Sustainable land-use patterns

We, cities and towns, recognise the importance of effective land-use and development planning policies by our local authorities which embrace the strategic environmental assessment of all plans. We should take advantage of the scope for providing efficient public transport and energy which higher densities offer, while maintaining the human scale of development. In both undertaking urban renewal programmes in inner urban areas and in planning new suburbs we seek a mix of functions so as to reduce the need for mobility. Notions of equitable regional interdependency should enable us to balance the flows between city and countryside and prevent cities from merely exploiting the resources of surrounding areas.

I.9 Sustainable urban mobility patterns

We, cities and towns, shall strive to improve accessibility and sustain social welfare and urban lifestyles with less transport. We know that it is imperative for a sustainable city to reduce enforced mobility and stop promoting and supporting the unnecessary use of motorised vehicles. We shall give priority to ecologically sound means of transport (in particular walking, cycling, public transport) and make a combination of these means the centre of our planning efforts. Motorised individual means of urban transport ought to have the subsidiary function of facilitating access to local services and maintaining the economic activity of the city.

I.10 Responsibility for the global climate

We, cities and towns, understand that the significant risks posed by global warming to the natural and built environments and to future human generations require a response sufficient to stabilise and then to reduce emissions of greenhouse gases into the atmosphere as soon as possible. It is equally important to protect global biomass resources, such as forests and phytoplankton, which play an essential role in the earth's carbon cycle. The abatement of fossil fuel emissions will require policies and initiatives based on a thorough understanding of the alternatives and of the urban environment as an energy system. The only sustainable alternatives are renewable energy sources.

I.11 Prevention of ecosystems toxification

We, cities and towns, are aware that more and more toxic and harmful substances are released into the air, water, soil, food, and are thereby becoming a growing threat to human health and the ecosystems. We will undertake every effort to see that further pollution is stopped and prevented at source.

I.12 Local self-governance as a pre-condition

We, cities and towns, are confident that we have the strength, the knowledge and the creative potential to develop sustainable ways of living and to design and manage our cities towards sustainability. As democratically elected representatives of our local communities we are ready to take responsibility for the task of re-organising our cities and towns for sustainability. The extent to which cities and towns are able to rise to this challenge depends upon their being given rights to local self-governance, according to the principle of subsidiarity. It is essential that sufficient powers are left at the local level and that local authorities are given a solid financial base.

I.13 Citizens as key actors and the involvement of the community

We, cities and towns pledge to meet the mandate given by Agenda 21, the key document approved at the Earth Summit in Rio de Janeiro, to work with all sectors of our communities – citizens, businesses, interest groups – when developing our Local Agenda 21 plans. We recognise the call in the European Union's Fifth Environmental Action Programme "Towards Sustainability" for the responsibility for the implementation of the programme to be shared among all sectors of the community. Therefore, we will base our work on co-operation between all actors involved. We shall ensure that all citizens and interested groups have access to information and are able to participate in local decision-making processes. We will seek opportunities for education and training for sustainability, not only for the general population, but for both elected representatives and officials in local government.

I.14 Instruments and tools for urban management towards sustainability

We, cities and towns, pledge to use the political and technical instruments and tools available for an ecosystem approach to urban management. We shall take advantage of a wide range of instruments including those for collecting and processing environmental data; environmental planning; regulatory, economic, and communication instruments such as directives, taxes and fees; and mechanisms for awareness raising including public participation. We seek to establish new environmental budgeting systems which allow for the management of our natural resources as economically as our artificial resource, "money."

We know that we must base our policymaking and controlling efforts, in particular our environmental monitoring, auditing, impact assessment, accounting, balancing and reporting systems, on different types of indicators, including those of urban environmental quality, urban flows, urban patterns, and, most importantly, indicators of an urban systems sustainability.

We, cities and towns, recognise that a whole range of policies and activities yielding positive ecological consequences have already been successfully applied in many cities through Europe. However, while these instruments are valuable tools for reducing the pace and pressure of unsustainability, they do not in and of themselves reverse society's unsustainable direction. Still, with this strong existing ecological base, the cities are in an excellent position to take the threshold step of integrating these policies and activities into the governance process for managing local urban economies through a comprehensive sustainability process. In this process we are called on to develop our own strategies, try them out in practice and share our experiences.

Part II: The European Sustainable Cities and Towns Campaign

We, European cities and towns, signatories of this charter, shall move forward together towards sustainability in a process of learning from experience and successful local examples. We shall encourage each other to establish long-term local action plans (Local Agendas 21), thereby strengthening inter-authority co-operation, and relating this process to the European Union's actions in the field of the urban environment.

We hereby initiate The European Sustainable Cities and Towns Campaign to encourage and support cities and towns in working towards sustainability. The initial phase of this Campaign shall be for a two-year period, after which progress shall be assessed at a Second European Conference on Sustainable Cities and Towns to be held in 1996. We invite every local authority, whether city, town or county and any European network of local authorities to join the Campaign by adopting and signing this Charter. We request all the major local authority networks in Europe to undertake the coordination of the Campaign. A Co-ordinating Committee shall be established of representatives of these networks. Arrangements will be made for those local authorities which are not members of any network.

We foresee the principal activities of the Campaign to be to:

- facilitate mutual support between European cities and towns in the design, development and implementation of policies towards sustainability;
- collect and disseminate information on good examples at the local level;
- promote the principle of sustainability in other local authorities;
- recruit further signatories to the Charter;
- organise an annual "Sustainable City Award";
- formulate policy recommendations to the European Commission;
- provide input to the Sustainable Cities Reports of the Urban Environment Expert Group;
- support local policymakers in implementing appropriate recommendations and legislation from the European Union;
- edit a Campaign newsletter.

These activities will require the establishment of a Campaign Co-ordination.

We shall invite other organisations to actively support the Campaign.

Part III: Engaging in The Local Agenda 21 Processes: Local Action Plans Towards Sustainability

We, European cities and towns, signatories of this Charter, pledge by signing this Charter and joining the European Sustainable Cities and Towns Campaign that we will seek to achieve a consensus within our communities on a Local Agenda 21 by the end of 1996. This will meet the mandate established by Chapter 28 of Agenda 21 as agreed at the Earth Summit in Rio in June 1992. By means of our individual local action plans we shall contribute to the implementation of the European Union's Fifth Environmental Action Programme "Towards Sustainability." The Local Agenda 21 processes shall be developed on the basis of Part One of this Charter.

We propose that the process of preparing a local action plan should include the following stages:

- recognition of the existing planning and financial frameworks as well as other plans and programmes;
- the systematic identification, by means of extensive public consultation, of problems and their causes;

- the prioritisation of tasks to address identified problems;
- the creation of a vision for a sustainable community through a participatory process involving all sectors of the community;
- the consideration and assessment of alternative strategic options;
- the establishment of a long-term local action plan towards sustainability which includes measurable targets;
- the programming of the implementation of the plan including the preparation of a timetable and statement of allocation of responsibilities among the partners;
- the establishment of systems and procedures for monitoring and reporting on the implementation of the plan.

We will need to review whether the internal arrangements of our local authorities are appropriate and efficient to allow the development of the Local Agenda 21 processes, including long-term local action plans towards sustainability. Efforts may be needed to improve the capacity of the organisation which will include reviewing the political arrangements, administrative procedures, corporate and inter-disciplinary working, human resources available and inter-authority cooperation including associations and networks.

Signed in Aalborg, Denmark, 27 May 1994

Appendix B

EMERALD CITY: A ROLEPLAYING SUSTAINABILITY GAME

(Designed by Richard S. Levine, Ernest J. Yanarella and Dennis Marshall)

Three Players

Choose one from each category.

1. Industrial/commercial:

    Land developer
    Construction firm executive
    Department Store Chain Franchiser

2. Environmental:

    Environmentalist
    Historic preservationist
    Conservationist

3. Community:

    Social justice activist
    Minority business owner
    Minority city council member

Core Values

Industrial/commercial

1. Maximize profit
2. Promote growth and expansion
3. Exert control over the policy process

Environmental

1. Protect natural wilderness
2. Improve environment/eliminate environmental burdens and threats
3. Expand nonintrusive wilderness recreation areas

Community

1. Defend and preserve integrity of community/neighborhood
2. Promote social equity/community welfare
3. Enhance opportunities (employment, education, services, etc.) for local inhabitants

Site and Problem [Example]

A brownfield site near downtown experiencing deteriorating conditions (slums, crime, drugs and other associated problems). Complicated by buried toxic wastes seeping out of ground from an abandoned factory site. In many respects, ideally located in terms of proximity to central local markets, potential labor force and interstate transportation. Neighborhood populated predominantly by poor African-Americans living in substandard housing partly owned by absentee landlords. Historically left undeveloped due to local problems compounding one another.

Problem: the industrial player wishes to buy a large tract of the land to develop it.

Scenario A: Individual self-interest

The first scenario begins with assumptions and rules of play common to typical urban conflicts. The players enter the game with a clear sense of their individual self-interest and attachment to their core values; and each seeks to play the game with single-minded determination by advancing those self-interests and maximizing those core values without regard for the interests or well-being of the other players. External resources (such as those described below) are brought into play only to maximize competitive self-advantage or (given the rules of play) to bring the game to a stalemate. The game is a classic zero-sum game (where one player's gain is another player's loss).

Problem and opportunity

Industrialist: need for expansion; site viewed as unused land ripe for development.

Environmentalist: environmental cleanup/improvement; site as location for an inner-city park.

Community activist: a vital neighborhood riddled with slum conditions; site as target for community improvement.

First iteration: each game participant plays out the trajectory of his agenda until the interests of the other(s) impede and then obstruct the pursuit and achievement of that design; outcome: STALEMATE.

Second iteration: players negotiate with one another, seeking to find basis for an agreement; since each player has five power units and any solution to the problem requires 12 power units, bargaining among the players necessarily involves some compromise among the individual positions and tradeoffs among core values of various players; if no bargain is struck after this iteration, players try to negotiate one more time; likely outcome: environmentalist – no park, some green space; industrialist – partial development negotiated, but must subsidize some housing at site; community activist – infill housing; upgrade of existing stock and some infrastructure improvements.

Consequences

First play-through: STALEMATE, no new benefits to any party.

Second play-through: COMPROMISE at the expense of core values to all parties; result deemed a failure by each player's constituencies.

In light of the discussion of the agreement and its implications, if one player concludes that the NEGOTIATED AGREEMENT has so badly injured that player's core values, that player may push the "externality button," ending the game by removing it from the local context and taking it to a higher (city, state or national) level.

Exemplary externalities

- Developer: appeals to mayor/city council
- Environmentalist: files suit on basis of state or federal environmental regulations
- Activist: organizes mass demonstration with picketing and civil disobedience

Threat of use of externality button triggers one more round of negotiation at the local level.

Scenario B: Veil of ignorance

Operating under the veil of ignorance of their roles or core values, the players begin to compromise the maximum agenda of these different roles and yield some of the resources and options of each to the other roles. After several rounds of compromise, the players have agreed on what appears a more or less balanced agenda that equalizes the advantages and disadvantages among the conflicting goals. In effect, this becomes the least common denominator or solution where each role played receives a good deal less than he/she wants and gains some advantage that he/she did not initially have.

Basis: enlightened self-interest resolution, not solution; least common denominator.

Veil of Ignorance: Players enter the game without foreknowledge of what role/core values they will be assigned after the outcome is reached.

Only after outcome [most likely a COMPROMISE] is reached do players push the random button, which arbitrarily assigns them a role/core values.

The COMPROMISE outcome is evaluated against its costs to the core values of each player.

Externality button suspended, put out of play.

Scenario C: Sustainable commonwealth

Players bring new resources to the table, but only if they can be shown to offer some benefits to the other players. Previously, this move or rule of play was not permitted because: in the competitive, self-interest scenario (A), players brought in external resources either to gain a competitive advantage over the other players or to stalemate the game, while in the compromise, lowest common denominator scenario (B), players were moved out of ignorance to trade down and balance existing resources to roughly equalize gains and losses among the players.

In the sustainable commonwealth scenario (C), a distinctly different game – the sustainability game – evolves. Realizing the inadequacies of confrontation and compromise and governed by new rules of play, the players are inclined to dig more deeply into their individual resources to experiment with strategies that may not immediately yield maximum returns or paybacks but, by offering advantages and incentives to other players as well as themselves, encourage the other players similarly to introduce cooperative resources. With an increasing variety of complex resources to choose from, the players embark upon a multiple building form of play. Instead of competing for the adoption of models they favor, they work together to try to increase the value of their models by making them fit with other players' models. Soon, as the resources (political, economic, social) introduced into the game begin to increase, each of the players' core values are satisfied and the commonwealth increases as well. The richer the collaborative construct negotiated by the players, the more outside resources are attracted to the game. The competition for resources is replaced by the synergy of players with different resources to contribute.

The game is driven toward BALANCE and EQUITY: equity between the players who are yet able to realize their goals, equity between the particular site and the city, equity between the city and the environment and equity between the environment and the future. This is the Sustainable Commonwealth scenario, the true Sustainability Game.

Basis: diversity of interests grounded in sustainability framework to greatest common denominator enlarging the pie from zero-sum to non-zero-sum game.

Rules of Play

1. Overall sustainability is nonnegotiable
2. No stalemate
3. No compromise of core values of any player
4. If the externality (outside resource) button is pushed, the player doing so must demonstrate to (and convince) the other two players that they will benefit from this new resource put on the table
5. Problems (negative externalities) may not be exported to others or to the future

Players thus negotiate upward, using new resources introduced as provocations to enlarge the pie and to devise more creative solutions to a site that takes on the characteristics of a commons.

Sustainability enhanced by tendencies toward greatest common denominator, pursuit of the common weal (will) and necessity of seeking balancing and turning negative externalities into assets since they must be overcome within the balance-seeking process and not cast off onto others (people, regions, etc.) or to future generations.

REFERENCES

Ackerman, Bruce and William T. Hassler. 1981. *Clean Coal/Dirty Air: How the Clean Air Act Became a Multi-billion Dollar Bailout for the High-sulfur Coal Producers*. New Haven, CT: Yale University Press.

Alberta First. "Okotoks Profile." http://albertacommunityprofiles.com/Profile/Okotoks (accessed 17 June 2011).

American Coalition for Clean Coal Electricity website. http://www.cleancoalusa.org/ (last accessed 10 August 2009).

American Forum for Global Education. 1995. "Sustainable Community Network Case Studies: Chattanooga, TN." http://chicagoconservationcorps.org/blog/wp-content/uploads1/2009/07/S06%20Chattanooga.pdf (accessed 24 June 2011).

Amin, Samir. 1976. *Unequal Development*. New York: Monthly Review Press.

Anderson, Lori and Brett Hodson. 2002. "An Introduction to Healthy Communities and the Health Okotoks Project." Paper prepared for the Elected Officials Conference, Edmonton, Alberta, 20 June 2002.

Arendt, Hannah. 1958. *The Human Condition*. Chicago: University of Chicago Press.

Associated Press. 1995. "Chattanooga Reborn as Scenic City." *Augusta Chronicle*, 22 May. http://chronicle.augusta.com/stories/2005/05/22/liv_453920.shtml (accessed 24 June 2011).

AtKisson, Alan. 1990. "The Utility of Global Thinking." *In Context* 25 (Spring): 55.

Austin, David and Terry Dinan. 2005. "Clearing the Air: The Costs and Consequences of Higher CAFÉ Standards and Increased Gasoline Taxes." *Journal of Environmental Economics and Management* 50: 562–82.

Bach, David and Hillary Rossner. 2008. *Go Green, Live Rich: 50 Simple Ways to Save the Earth and Get Rich Trying*. New York: Broadway Books.

Bailes, John and Aaron Mesh. 2005. "The Lost World: Former Downtown Residents Say Their City was Stolen by Development – They Want it Back." *Pulse: Chattanooga's Weekly Alternative*.

Bailey, Blythe. 2009. "Planning for an Urban Future is All about Connections." *Pulse: Chattanooga's Weekly Alternative*, 29 April.

Balfour, Richard and Eileen McAdam Keenan. 2007. *Strategic Sustainability Planning: A Civil Defense Manual for Cultural Survival*. Vancouver, BC: Old City Foundation Press.

Ballance, Cindy. 2001. "Council Sees Sustainable Okotoks Concepts Come Alive through Design." *Western Wheel*, 24 October.

_______. 2004. "Okotoks' Municipal Manager Resigns Post." *Western Wheel*, 29 September.

Barlow, John. 2006. "Change is a Comin'." *Western Wheel*, 24 October.

_______. 2010. "Water Options are Drying Up." *Western Wheel*, 6 January.

Barnett, Jonathan, F. Kaid Benfield, Paul Famer, Shelley Poticha, Robert Yaro and Armando Carbonell. 2007. *Smart Growth in a Changing World*. Washington, DC: American Planning Association.

Bartling, Hugh, and Donald Ferris. 1997. "Chattanooga: Is This Sustainable?" PS 776 seminar hyperlink paper, University of Kentucky. http://www.uky.edu/Classes/PS/776/Projects/chattanooga/chattnga.html (accessed 10 August 2009).

Beatley, Timothy, David J. Brower and Ruth Knack. 1993. "Sustainability Comes to Main Street: Principles to Live – and Play – By." *Planning* 59.5: 16–19.

Beckwith, Richard. 2006. "What's Next: Shannon May." WorldChanging.com, 28 December. http://www.worldchanging.com/archives/005681.html (accessed 30 June 2010).

Bell, Daniel. 1960. *The End of Ideology: On the Exhaustion of Political Ideas in the Fifties*. New York: Free Press.

Benyus, Janine M. 2002. *Biomimicry: Innovation Inspired by Nature*. New York: Harper Perennial.

Berelowitz, Lance. 2005. *Dream City: Vancouver and the Global Imagination*. Vancouver, BC: Douglas & McIntyre Publishers, Inc.

Berry, Aylssa. 2005. Interview, in the town of Okotoks, Alberta. 10 August.

Berry, Ian. 2007. "EPA, Companies at Odds over Creek Cleanup." *Chattanooga Times Free Press*, 16 May.

Berry, Wendell. 1989. "The Futility of Global Thinking. *Harper's* (September): 16–22.

_______. 2001. *Life is a Miracle: An Essay Against Modern Superstitution*. Berkeley, CA: Counterpoint Press.

Boniface, Russell. 2009. "Sustainable Design Assessment Team Aims for Green in Tennessee." *AIArchitect This Week*, 8 May. http://info.aia.org/aiarchitect/thisweek09/0508/0508p_sdat.cfm (accessed 17 June 2011).

Bookchin, Murray. 1974. *The Limits of the City*. New York: Harper & Row.

_______. 1996. *The Limits of the City*. Montreal, QC: Black Rose Books.

Boyer, M. C. 1996. *Cybercities: Visual Perception in the Age of Electronic Communications*. New York: Princeton Architectural Press.

Brooke, James. 1988. "Waste Dumpers Turning to West Africa." *New York Times*, 17 July.

Brown, Lester. 1981. *Building the Sustainable Society*. New York: W. W. Norton & Company.

Brugmann, Jeb. 1996. "Planning for Sustainability at the Local Governmental Level." *Environmental Impact Assessment Review* 16: 363–79.

Brunn, Stanley D. and Ernest J. Yanarella. 1987. "Toward a Humanistic Political Geography." *Studies in Comparative International Development* 22.2: 1–49.

Bryner, G. C. 2001. *Gaia's Wager: Environmental Movements and the Challenge of Sustainability*. New York: Rowman & Littlefield, Publishers.

Buderer, Jeff. 2006a. "April Streeter Reports on Big Trouble in Rural China." *EcoLiving Solutions*, 8 July. http://green.onevillage.tv/?p=10 (accessed 7 June 2007).

_______. 2006b. "Bill McDonough's Huangbaiyu Ecovillage Risks Failure by Overlooking the Needs of Local People." *Ecoliving Solutions*, 9 July. http://green.onevillage.tv/?p=44 (accessed 17 June 2011).

Bukatman, Scott. 1993. *Terminal Identity: the Virtual Subject in Postmodern Science Fiction*. Durham, NC: Duke University Press.

Campbell, C. J. and J. H. Laherrere. 1998. "The End of Cheap Oil." *Scientific American* 278 (March): 78–83.

Canadian Mortgage and Housing Corporation. "Sustainable Community Design Demonstration in Okotoks, Alberta: Testing Consumer Receptivity." *Research Highlights*, Socio-Economic Series 1, December 2002.

Carroll, Beverly A. 2009a. "More Voices on Gentrification." *Pulse: Chattanooga's Weekly Alternative*, 12 August.

______. 2009b. "Social Equity and Chattanooga's Renewal: Whose Backyard is It Anyway?" *Pulse: Chattanooga's Weekly Alternative*, 15 October.

Carson, Rachel. 1994. *Silent Spring* (reprint). New York: Ballantine Books.

Castells, M. 1989. *The Informational City: Information Technology, Economic Restructuring and the Urban-Regional Process.* Cambridge, MA: Blackwell Publishers.

______. 1999. "The Information City is a Dual City: Can It Be Reversed?" In D. A. Schon, B. Sanyal and W. C. Mitchell (eds), *High Technology and Low-Income Communities: Prospects for the Positive Use of Advanced Information Technology*, 25–42. Cambridge, MA: MIT Press.

Ceraso, Karen. 1999. "Cleaner, Greener Chattanooga: After a Stunning Environmental Turnaround, Can Chattanooga Become a True Resident-driven Sustainable City?" *Shelterforce Online* 103. http://www.nhi.org/online/issues/103/ceraso.html (accessed 30 January 2010).

Chaoui, Lamia, E. Leman and Z. Rufei. 2009. "Urban Trends and Policies in China." OECD Regional Development Working Papers 2009/1. doi:10.1787/225205036417.

Charlie Rose Show transcript. 2006. "Environmentalist Amory Lovins from Rocky Mount Institute on Energy Alternatives." PBS, 28 November. http://video.google.com/videoplay?docid=4569577556800822039# (accessed 17 June 2011).

Chattanooga Chamber of Commerce. 1996. "Press Release: Chattanooga Strategy for Sustainable Development." Chamber of Commerce press release.

______. 2007. "Alstom Announces New Facility, 350 Jobs." Chamber of Commerce news posting, 14 December.

______. 2008. "Volkswagen chooses Chattanooga's Enterprise South Industrial Park." Chamber of Commerce news posting, 15 July.

Chattanooga Green community profile. 2009. http://community.greenerchattanooga.com/profile/BradMcAllister?xg_source=activity (accessed 24 June 2011).

Chattanooga Green Committee. 2008. "Chattanooga's Climate Action Plan – Interim Report." Report produced for the Chattanooga City Government, June. http://www.chattanooga.gov/Chatt_Green_Interim_Report.pdf (accessed 1 July 2011).

______. 2009. "Chattanooga's Climate Action Plan – Final Report." Report produced for the Chattanooga City Government, January. http://www.chattanooga.gov/Final_CAP_adopted.pdf (accessed 1 July 2011).

Chattanooga Green website. http://www.chattanooga.gov/ChattanoogaGreen_ChattanoogaGreen.htm (last accessed 28 May 2010).

Chattanooga Institute website. http://www.csc2.org (last accessed 4 April 2000).

Chattanooga Neighborhood Enterprise website. http://www.cneinc.org (last accessed 30 January 2010).

Chevalier, Jessica. 2008. "Chattanooga Going Green." *CityScope Magazine*, October.

China-US Center for Sustainable Development. 2006. "The Village of Huangbaiyu: Exploring Sustainable Design." http://gilhardwick.com/reports/Environmental%20Management%20Systems/Check%20Your%20Success%20-%20A%20Guide%20to%20Developing%20Indicators.pdf (accessed 17 June 2011).

Citizens for a Sustainable Okotoks Facebook page. http://www.facebook.com/board.php?uid=4561327683 (last accessed 17 June 2010).

Citizens for a Sustainable Okotoks. 2009. "Citizens for a Sustainable Okotoks Submission to the CRP." 24 May. http://nocalgaryveto.wordpress/2009/05/24/cso_submission/ (website discontinued – last accessed 17 June 2010).

Cohen, Mark. 2008. "Extra Cleanup Suggested for Chattanooga Creek." *Chattanooga Times Free Press*, 5 May.

Colgan, Charles S. 1997. "'Sustainable Development' and Economic Development Policy: Lessons from Canada." *Economic Development Quarterly* 11: 123–37.

Connolly, William W. 1983. *Terms of Political Discourse*. Oxford: Martin Robertson.

Daly, Herman. 1980. *Economics, Ecology, Ethics*. San Francisco: Freeman Press.

_______. 1997. *Beyond Growth: The Economics of Sustainable Development*. Boston: Beacon Press.

Daly, Herman and John B. Cobb, Jr. 1989. *For the Common Good: Redirecting the Economy toward Community, the Environment, and a Sustainable Future*. Boston: Beacon Press.

_______. 1994. *For the Common Good: Redirecting the Economy toward Community, the Environment, and a Sustainable Future*. Boston: Beacon Press.

Deffeyes, Kenneth S. 2005. *Beyond Oil: The View from Hubbert's Peak*. New York: Hill & Wang.

_______. 2008. *Hubbert's Peak: the Impending World Oil Shortage*. Princeton, NJ: Princeton University Press.

DeKay, Mark, and Tracy Moir-McClean with Richar Rothman. 2003. *Greencenter: Planning for Environmental Quality in Downtown Chattanooga*. A report to the Chattanooga Downtown Planning and Design Center. Chattanooga, TN: Green Vision Studio.

Department of Urban Affairs, West Virginia University. n.d. "All Aboard for Chattanooga Creek, Chattanooga Tennessee." In *Check Your Success: A Guide to Developing Indictors for Community Based Environmental Projects*. Morgantown, WV: Department of Urban Affairs, West Virginia University. http://www.uap.vt.edu/checkyoursuccess/cs3.html (website discontinued – last accessed 28 May 2010).

Dickens, Charles. [1859] 1997. *A Tale of Two Cities*. New York: Signet Classics.

Doheny-Farina, S. 1996. *The Wired Neighborhood*. New Haven, CT: Yale University Press.

Domhoff, G. William. 2005. "Power at the Local Level: Growth Coalition Theory." whorulesamerica.net. http://sociology.ucsc.edu/whorulesamerica/local/growth_coalition_theory.html (accessed 10 August 2009).

Drake Landing Solar Community website. http://www.dlsc.ca (last accessed 17 June 2010).

Dumreicher, Heidi. 2006. "Final Report of the SUCCESS Project of the European Commission 5th Framework Programme, 2002–2005." (Project No. ECA4-CT-2002-10007). Vienna: Oikodrom – The Vienna Institute for Urban Sustainability.

Dumreicher, Heidi and Richard S. Levine. 1995. *Stadthugel Westbahnhof EinXos/probe*. Vienna: Oikodrom – The Vienna Institute for Urban Sustainability.

_______. 1996. *Stadhugel Wien Weslbahnhof Zweitel. Teil Die Dreidimensionale Stadt Strassen, Plarze, Menschen, Arbrit*. Vienna: Oikodrom – The Vienna Institute for Urban Sustainability.

Dumreicher, Heidi and Veronika Prañdl-Zika. 2008. "Chinese Villages and Their Sustainable Future" (Special Issue on SUCCESS research and findings). *Journal of Environmental Management* 87.2 (April).

Dumreicher, Heidi, Bettina Kolb, Richard S. Levine, Lu Hongyi and Ina Ivanceanu. 2005. "Sustainability Scenarios for Spatial Planning: Ideal or Real?" *CORP 2005 & Geomultimedia 05*, February 22–5.

Dumreicher, Heidi, Richard S. Levine, Ernest J. Yanarella and Taghi Radmard. 2000. "Generating Models of Urban Sustainability: Vienna's Westbahnhof Sustainable Hilltown." In Katie Williams, Elizabeth Burton and Mike Jenks (eds), *Achieving Sustainable Urban Form*, 288–98, 379. London: Routledge.

Ecoworks. 2006. "Featured Project: Chinese Sustainable Village Prototype." http://www.ecoworksfoundation.org/projectsVillage.html (accessed 7 June 2007).

_______. 2007. "Chinese Sustainable Village Prototype Update." http://www.ecoworksfoundation.org/projectsVillage_Update.html (accessed 7 December 2007).

Egelko, Bob. 2010. "Pleasant Housing Cap Violates Law, Judge Says." *San Francisco Chronicle*, 16 March: C1.

Ehrenfeld, David. 1981. *The Arrogance of Humanism*. New York: Oxford University Press.

Eichenthal, David and Tracy Winderknecht. 2008. *Chattanooga, Tennessee: A Restoring Prosperity Case Study*. Washington, DC: Brookings Institution.

Energy Solutions Alberta. "Sustainable Okotoks." http://www.energysolutionsalberta.com/?V_DOC_ID=966 (website discontinued – last accessed 17 June 2010).

Environmental City Task Force. 2001. "Environmental Vision for Chattanooga." *ECTF* website, 15 February. http://www.envirocity.org (website discontinued).

European Commission. 1994. "The Charter of European Cities and Towns Towards Sustainability (Aalborg Charter)." http://ec.europa.eu/environment/urban/pdf/aalborg_charter.pdf (accessed 19 June 2011).

Exxon Mobil. 2009. "Save the Tiger Fund: Brochure." http://exxonmobil.com/Corporate/Files/Corporate/STF_brochure.pdf (accessed 10 August 2009).

Faculty of Environmental Design. n.d. "Sustainable Okotoks: Sustainable Design Demonstration Research Program." Calgary, AB: College of Architecture, University of Calgary. http://www.architecture.ca/okotoks/index.htm (website discontinued – last accessed 17 June 2010).

Fan, Cindy C. 2006. "China's Eleventh Five-Year Plan (2006–2010): From 'Getting Rich First' to Common Prosperity." *Eurasian Geography and Economics* 47.6: 708–21.

Fanon, Frantz. 2005. *The Wretched of the Earth*. New York: Grove Press.

Filion, Pierre, Heidi Hoernig, Trudi Bunting and Gary Sands. 2004. "The Successful Few: Healthy Downtowns of Small Metropolitan Regions." *Journal of the American Planning Association* 70.3: 328–43.

Fitzgerald, Joan. 2010. *Emerald Cities: Urban Sustainability and Economic Development*. New York: Oxford University Press.

Flessner, Dave. 2007a. "Audit Cites CNE 'Poor Management'." *Chattanooga Times Free Press*, 27 January.

_______. 2007b. "Internal Audits Shake Up Staff at City Hall." *Chattanooga Times Free Press*, 5 September.

Flessner, Dave and Pam Sohn. 2008. "Chattanooga Cleans Up for a Better Quality of Life." *Chattanooga Times Free Press*, 27 April. http://www.timesfreepress.com/news/2008/apr/27/chattanooga-looking-green-image-future----again/ (accessed 25 July 2011).

Foster, John Bellamy and Fred Magdoff. 2009. *The Great Financial Crisis: Causes and Consequences*. New York: Monthly Press.

Frampton, Kenneth. 1983. "Towards a Critical Regionalism: Six Points for an Architecture of Resistance." In Hal Foster (ed.), *The Anti-Aesthetic: Essays on Postmodern Culture*, 16–30. Port Townsend, WA: Bay Press.

Friedman, Thomas. 2008. *Hot, Flat, and Crowded: Why We Need a Green Revolution – and How It Can Renew America*. New York: Farrar, Straus and Giroux.

Frierson, James. 2003. Interview at Chattanooga City Hall, April.

Fukuyama, Francis. 1992. *The End of History and the Last Man*. New York: Harper Perennial.

Gajjar, Kishan N., Will Peace, William T. Perks and Mary-Ellen Tyler. n.d. *Sustainable Community Design*. CD-ROM.

Gallie, W. B. 1956. "Essentially Contested Concepts." *Aristotelian Society* 56: 167–98.

Garreau, Joel. 2006. *Radical Evolution: The Promise and Peril of Enhancing Our Minds, Our Bodies – and What It Means to Be Human*. New York: Broadway Books.

Germaine, L., Sandra J. Phifer, Judy A. Nixon and Marilyn Wood. 1998. *Rubrics – Handbook for Construction and Use*. Lancaster, PA: Technomic Publishing.

Girardet, Herbert. 2006. "Which Way China?" *ChinaDialogue*, 2 October. http://www.chinadialogue.net/article/show/single/en/297-Which-way-China- (accessed 17 June 2011).

Gibson, Amy Kay. 1998. "Sustaining the Treadmill of Production: Chattanooga and the Social Construction of a Solution." PhD dissertation, University of Tennessee.

Glick, Daniel. 1996. "Cinderella Story." *National Wildlife Magazine*, 1 February. http://www.nwf.org/News-and-Magazines/National-Wildlife/News-and-Views/Archives/1996/Cinderella-Story.aspx (accessed 28 May 2010).

Goldsmith A. 2007. "Here's an Idea That's Not Quite Ripe." *FastCompany.com*, 17 December. http://www.fastcompany.com/magazine/11/cdu.html (accessed 17 June 2011).

Gorz, Andre. 1972. *A Strategy for Labour: A Radical Proposal.* Boston: Beacon Press.

Graham, S and S. Marvin. 1996. *Telecommunications and the City: Electronic Spaces.* New York: Routledge.

Gramsci, Antonio. 1971. *Selections from the Prison Notebooks.* Trans. Quintin Hoare and ed. Geoffrey Nowell Smith. New York: International Publishers.

Green, Molly. 2005. Interview in the town of Okotoks, Alberta, 10 August.

Greer, Jed and K. Bruno. 1997. *Greenwash: the Reality Behind Corporate Environmentalism.* New York: Apex Press.

Grunkemeyer, William and Myra Moss. n.d. "Key Concepts in Sustainable Development: The Chattanooga Model." *The Web Book of Regional Science.* http://rri.wvu.edu/WebBook/Grunkemeyer-Moss/contents.htm (accessed 17 June 2011).

Guttormson, Kim. 2008. "Okotoks Mayor Praises Proposal for Express Buses." *Calgary Herald*, 30 July. B4.

HABICO Planning + Architecture Ltd. "Town of Okotoks Downtown Revitalization: Public Space Design Guidelines." http://www.habico.com/Okotoksbody.html (accessed 17 June 2010).

Hamrin, Robert D. 1983. *A Renewable Resource Economy.* New York: Praeger Publishers.

Haraway, D. J. 1989. "Manifesto for Cyborgs: Science, Technology, and Socialist Feminism in the 1980s." *Socialist Review* 80: 65–108.

Hartshorn, Truman A. and Susan M. Walcott. 2000. "The Three Georgias: Emerging Realignments at the Dawn of the New Millennium." *Southeastern Geographer* 41.2: 127–50.

Hashe, Janis. 2009. "Cover Story: What We Will Be." *Pulse: Chattanooga's Weekly Alternative*, 18 March.

Hauben, M and R. Hauben. 1997. *Netizens: On the History of Usenet and the Internet.* New York: IEEE Computer Society.

Hawken, Paul. 1994. *The Ecology of Commerce: a Declaration of Sustainability.* New York: Harper Business.

Hawken, Paul, Amory Lovins and Hunter L. Lovins. 1999. *Natural Capitalism: Creating the Next Industrial Revolution.* New York: Back Bay Books

Hays, Samuel P. 1969. *Conservation and the Gospel of Efficiency: The Progressive Conservation Movement.* New York: Atheneum Books.

Henderson, Hazel. 1981. "Thinking Globally, Acting Locally: Ethics for the Dawning Solar Age." In *The Politics of the Solar Age: Alternatives to Economics.* New York: Anchor Books. 355–405.

_______. 1988. "Thinking Globally, Acting Locally: Ethics for the Dawning Solar Age." In *The Politics of the Solar Age: Alternatives to Economics.* Indianapolis: Knowledge Systems, Inc.

Hess, David. 2005. "Case Studies of the Greening of Urban Transit: Chattanooga Area Regional Transportation Authority." Department of Science and Technology Studies, Rensselaer Polytechnic Institute. http://www.davidjhess.org/CARTA.pdf (accessed 17 June 2011).

Hightower, Cliff. 2009a. "Chattanooga: City Receives Suggestions toward Sustainability." *Chattanooga Times Free Press*, 16 January.

_______. 2009b. "Office of Sustainability on Track, Littlefield Says." *Chattanooga Times Free Press*, 30 November.

Hodson, Brett and Lori Anderson. 2002. "A Made-in-Alberta Healthy Community." *Plan Canada* 42: 18–19.

Hodson, Brett, Lori Anderson, Donna Anderson, Kim Raine-Travers, and Carey Shore. 2000."Evaluation of the Healthy Okotoks Project." Report prepared for the Alberta Health and Wellness Health Transition Fund, 30 June.

Holmgren, David. 2002. *Permaculture: Principles & Pathways Beyond Sustainability*. Hepburn, Vic.: Holmgren Design Services.

_______. 2009. *Future Scenarios: How Communities Can Adapt to Peak Oil and Climate Change*. White River Junction, VT: Chelsea Green Publishing.

Hopkins, Rob. 2008. *The Transition Handbook: From Oil Dependency to Local Resilience*. Tornes: Green Books.

Howard, Ebenezer. [1902] 1965. *Garden Cities of To-Morrow*. Cambridge, MA: MIT Press.

Hu, Binliang. 2008. "China's 'New Village' Strategy—Actual Progress and National Impact" Transcript. Speech before the Carnegie Endowment for International Peace, 31 January. Washington, DC: Federal News Service.

Huang, Daohen. 2009. "Doubts Mount over Our Country's First 'Carbon-free' City." *Beijing Today*, 1 January. http://bjtoday.ynet.com/article.jsp?oid=47850569 (accessed 23 February 2010).

Hubbert, M. King. 1962. "Energy Resources: a Report to the Committee on Natural Resources of the National Academy of Sciences – National Research Council." Washington, DC: National Research Council/National Academy of Sciences.

Hum, Sir Christopher. 2006. "China: The 11th Five-Year Plan and the Challenges of Development." Speech given at the KC Wu Memorial Lecture, Guild of International Bankers and Cass Business School, June.

Hvistendahl, Mara. 2007. "China Eco-Cities Update." WorldChanging.com, 22 August. http://www.worldchanging.com/archives/007153.html (accessed 7 December 2007).

IBRD/World Bank. 1997. *World Bank Atlas 1997*. Washington, DC: International Bank for Reconstruction and Development/World Bank.

Idemudia, Patience and Kole Shettima. 1994. "The World Bank Takes Control of UNCED's Environment Fund." In Kevin Danaher and Muhammad Yunus (eds), *50 Years Is Enough: The Case Against the World Bank and the International Monetary Fund*, 2–6. Boston: South End Press.

Intergovernmental Panel on Climate Change (IPCC). 2007. "Summary for Policymakers." In S. Solomon, D. Qin, M. Manning, Z. Chen, M. Marquis, K. B. Averyt, M. Tignor and H. L. Miller (eds), *Climate Change 2007: The Physical Science Basis. Contribution of Working Group I to the Fourth Assessment Report of the Intergovernmental Panel on Climate Change*, 1–17. New York: Cambridge University Press.

International Council for Local Environmental Initiatives website. http://www.iclei.org (last accessed 1 July 2011).

International Council for Local Environmental Initiatives (ICLEI).1997a. "ICLEI's Role in LA 21: A Report" http://www.iclei.org/iclei/la21.htm (website discontinued – last accessed 25 April 2001).

_______. 1997b. "Local Government Implementation of Agenda 21: A Report." http://www.iclei.org/la21/la21lgov.htm (website discontinued – last accessed 25 April 2001).

_______. n.d. "Case Studies on the Local Agenda 21 Process: Durban." http://www. iclei.org/csdcases/durban.htm (website discontinued – last accessed 25 April 2001).

_______. n.d. "Case Studies on the Local Agenda 21 Process: Hamilton-Wentworth." http://www.iclei.org/csdcases/hamwent.htm (website discontinued – last accessed 25 April 2001).

_______. n.d. "Case Studies on the Local Agenda 21 Process: Santos." http://www. iclei.org/csdcases/santos.htm (website discontinued – last accessed 25 April 2001).

_______. n.d. "Case Studies on the Local Agenda 21 Process: Troyan." http://www. iclei.org/csdcases/troyan.htm (website discontinued – last accessed 25 April 2001).

Jao, Joseph. 2005. "Healthy Growth of Cities." *China Daily*, 5 November. http://www.chinadaily.com.cn/english/doc/2005-11/05/content_491585.htm (website discontinued – last accessed 17 May 2007).

Jepson, W., R. Liggett and S. Friedman. 1996. "Virtual Modeling of Urban Environments." *Presence* 5.1.

Jones, Charisse. 2005. "A Whiff of Politics in Gotham's Garbage." *USA Today*. http://www.usatoday.com/news/nation/2005-07-19-nyc-garbage_x.htm (accessed 17 June 2011).

Joplin, John and Herbert Girardet. 1997. *Creating a Sustainable London*. London: Sustainable London Trust.

Kahn, Herman. 1976. *The Next Two Hundred Years*. New York: William Morrow & Company.

Kalicki, Jan H. and David L. Goldwyn. 2005. *Energy and Security: Toward a New Foreign Policy Strategy*. Baltimore, MD: Johns Hopkins University Press.

Kendrick, Martyn and Linda Moore. 1995. *Re-inventing Our Common Future: An Exploration into Community Sustainability*. Hamilton, ON: Eco Gateway Group.

Kibert, Carles J. 2007. *Sustainable Construction: Green Building Design and Delivery*. New York: John Wiley & Sons.

Kiesler, S. 1996. *Culture of the Internet*. Mahwah, NJ: Lawrence Erlbaum Associates.

King, J. L., R. E. Ginter and J. Pickering. 1996. "The Rise and Fall of Netville: the Saga of a Cyberspace Construction Boomtown in the Great Divide." In S. Kiesler (ed.), *Culture of the Internet*, 3–33. Mahwah, NJ: Lawrence Erlbaum Associates.

Knight, Patrick. 1996. "Sustainable Urban Planning in Santos, Brazil." International Development Research report. http://www.idrc.ca/books/reports/1996/03-01e.html (website discontinued – last accessed 16 May 2001).

Kraft, Michael. 1977. "Political Change and the Sustainable Society." In Dennis Pirages (ed.), *The Sustainable Society*, 173–96. New York: Praeger Publishers.

Krasner, Stephen. 1985. *Structural Conflict: The Third World Against Global Liberalism*. Berkeley, CA: University of California Press.

Krueger, R. and J. Agyeman. 2005. "Sustainability Schizophrenia or 'Actually Existing Sustainabilities?' Toward a Broader Understanding of the Politics and Promise of Local Sustainability in the US." *Geoforum* 36: 410–17.

Krugman, Paul. 2009. *The Return of Depression Economics and the Crisis of 2008*. New York: W. W. Norton & Company.

Kunstler, James Howard . 1994. *Geography of Nowhere: The Rise and Decline of America's Man-Made Landscape*. New York: Free Press.

_______. 2005. *The Long Emergency: Surviving the Converging Catastrophes of the Twenty-first Century*. New York: Atlantic Monthly Press.

_______. 2006. *The Long Emergency: Surviving the End of Oil, Climate Change, and Other Converging Catastrophes of the Twenty-first Century*. New York: Grove Press.

Lancaster, Robert. 2004. "Chattanooga, Tennessee and VISION 2000." In *The Uncertain Quest for Urban Sustainability: Case Studies of Local Sustainability in Comparative Perspective*, chapter 8. PhD dissertation, University of Kentucky.

Lanier, Jaron. 2010. *You Are Not a Gadget*. New York: Alfred A. Knopf.

Lawson, Colin W. 1985. "The Soviet Union in North-South Negotiations: Revealing Preferences." In Robert Cassen (ed.), *Soviet Interest in the Third World*, 177–91 Beverly Hills, CA: SAGE Publications.

Lenin, V. I. 1966. *Collected Works*. Volume 31. Moscow: Progress Publishers.

Lesle, Timothy. 2007. "Green Dreams." http://www.pbs.org/frontlineworld/fellows/green_dreams/ (accessed 17 June 2010).

Levine, Richard S. 1982. "The Coupled Pan Space Frame: A Structural Framework for Solar Conserving Buildings." In *Progress in Passive Solar Energy: Proceedings of the American Solar Energy Society*, Knoxville, TN: 7th National Passive Solar Conference.

_______. 1987. "The Future Medieval City." *Spacio e societa/Space & Society* 10: 18–24.

_______. 1989. "The Sustainable City: A Necessary Utopia." In Gisuseppa Saccaro Del Buffa and Arthur O. Lewis (eds), *Utopia and Modernity: Theory and Praxis in the Modern and Postmodern Era*, 157–73. Rome: Gangemi editore.

_______. 1992. "'Design' as the operative model for generating sustainable cities." Paper presented at the American Solar Energy Society Conference, Cocoa Beach, Florida, May.

_______. 1994. "The Sustainable City: The Scientific Design Study and the Sustainable City Game." Unpublished paper.

_______. 1996. "The Sustainability Engine© and the City." *Stadlpläne* 6: 1–44.

Levine, Richard S. and Ernest J. Yanarella. 1991. "The Once and Future Sustainable City." Paper presented at the International Solar Energy Society World Solar Congress, Denver, Colorado, August.

Levine, Richard S. and Taghi Radmard. 1990. "Sustainable City Implantations." Paper presented at the American Solar Energy Society Conference, Austin, Texas, 19–22 March.

_______. 1992. "The Two Step Process for Generating Sustainable Cities." Paper prepared for the Second International Ecological Cities Conference, Adelaide, Australia, 13 June.

Levine, Richard S., Ernest J. Yanarella, Heidi Dumreicher and Tiffany Broyles. 2000. "Beyond Sustainability Indicators: The Sustainable Area Budget." Paper presented at the Making Sustainable Development Visible: Indicators for Regional Development Conference, Graz, Austria, 13–15 November.

Levine, Richard S., Taghi Radmard and David Harper. 1991. "The Development of a Computer Aided Design Model for Generating Sustainable Cities." Paper presented at the ISES World Solar Congress, Denver, Colorado, August.

Levine, Richard S., Ernest J. Yanarella and Taghi Radmard. 1991. "The Development of an Interactive Computer-aided Design Model for Generating the Sustainable City." Paper prepared for the International Solar Energy Society, World Solar Congress, Denver, Colorado, 15 April.

Levine, Richard S., Michael T. Hughes, Casey Ryan Mather and Ernest J. Yanarella. 2008. "Generating Sustainable Towns from Chinese Villages: A Systems Modeling Approach." *Journal of Environmental Management*. 87.2: 305–16.

Levine, Richard S., Ernest J. Yanarella, Heidi Dumreicher and Tiffany Broyles. 2000. "Beyond Sustainability Indicators: The Sustainable Area Budget." Paper presented at the Making Sustainable Development Visible: Indicators for Regional Development Conference, Graz, Austria, 13–15 November.

Levine Richard S., Ernest J. Yanarella, Taghi Radmard and David Harper. 1991. "The Development of an Interactive Computer-aided Design Model for Generating the Sustainable City." Unpublished paper.

Levine, Richard S, Ernest J. Yanarella, Taghi Radmard and Heidi Dumreicher. 2003. "Sustainable Cities: a Strategy for a Post-terrorized World." *Terrain.org: A Journal of the Built & Natural Environments* 13 (Summer/Fall). http://www.terrain.org/articles/13/strategy.htm (accessed 17 June 2010).

Levine, Richard S., Michael T. Hughes, Casey Ryan Mather, Heidi Dumreicher and Lv Hongyi. 2007. "The Proto-Sustainable Village as Generator of the Future Chinese City." SUCCESS paper.

Limbird, Larry. 2003. "Sustainable Okotoks: Bringing the Sustainability Concept to Life." *AACIP Planning Digest* 7 (Spring): 2–7.

Littlefield, Ron. 2008. "Perspective: Chattanooga Won't Abandon Environmental Initiatives." *Chattanooga Times Free Press*, 16 November.

Logan, John R. 2008. *Urban China in Transition*. Malden, MA: Blackwell.

Logan, John R. and Harvey L. Molotch. 1987. *Urban Fortunes: The Political Economy of Place*. Berkeley, CA: University of California Press.

_______. 2007. *Urban Fortunes*. Berkeley, CA: University of California Press.

Lohmann, Larry. 1993. "Resisting Green Globalism." In Wolfgang Sachs (ed.), *Global Ecology: A New Arena of Political Conflict*, 159–69. Atlantic Heights, NJ: Zed Books.

Lovelock, James. 1979. *Gaia: A New Look at Life* on *Earth*. New York: Oxford University Press.

_______. 2007. *The Revenge of Gaia: Earth's Climate Crisis and the Fate of Humanity*. New York: Basic Books.

Lovins, Amory B. 1971. *Soft Energy Paths: Toward a Durable Peace*. New York: Ballinger Books.

_______. 1976. "Energy Strategy: The Road Not Taken?" *Foreign Affairs* 55 (October): 65–96.

_______. 1977. *Soft Energy Paths: Toward a Durable Peace*. New York: Ballinger Publishing Company.

Lukács, Georg. 1968. *History and Class Consciousness*. London: Merlin Press.

Lukasik, Lynda. 1996. "Letter to the Editor: It's the Big Picture." *Hamilton Spectator*, 7 September: A12.

Luke, Tim. 1999. "Simulated Sovereignty: Telematic Territoriality: Political Economy of Cyberspace." In M. Featherstone and S. Lash (eds), *Spaces of Culture: City Nation World*, 27–48. London: Sage.

Lynch, Maureen. 2005. "Tri-Community Watershed Initiative: Tows of Black Diamond, Turner Valley and Okotoks, Alberta Promoting Sustainable Behaviour in Watersheds and Communities." http://www.fs.fed.us/rm/pubs/rmrs_p042/rmrs_p042_462_468.pdf (accessed 17 June 2011).

Madison, Cathy. 1997. "Chattanooga, Tennessee: The Sustainable Blue-Collar Town." *Utne Reader*, May–June. http://www.utne.com/Politics/Americas-Most-Enlightened-Towns-Chattanooga-Tennessee.aspx (accessed 24 June 2011).

Mallet, James S. "Municipal Powers, Land Use Planning, and the Environment: Understanding the Public's Role." Edmonton, AB: Environmental Law Centre.

Markusoff, Jason. 2009. "Capping Town's Size Sparks Controversy in Okotoks." *Edmonton Journal*, 26 May: B6.

Marschalek, Ilse. 2008. "The Concept of Participatory Local Sustainability Projects in Seven Chinese Villages." *Journal of Environmental Management* 87.2: 226–35.

Marshall, Alex. "Chattanooga in Metropolis." Irresponsible Journalism: The Blog of Josiah Q. Roe. http://chattablogs.com/quintus/archives/029976.htm (website discontinued – last accessed 28 May 2010).

Marston, J. Ed. 2004. "Chattanooga's Changes: Public-private Partnerships are Revitalizing the Community." *Urban Land*, June: 38–41.

Martens, Susan. 2006. "Public Participation with Chinese Characteristics: Citizen Consumers in China's Environmental Management." *Environmental Politics* 15.2: 211–30.

May, Shannon. 2007. "A Sino-U.S. Sham." *Far Eastern Economic Review* 170.3: 57–60.

_______. 2008a. "Econ with Eco: Hope and Hazard in Rural China." *Far East Economic Review* 171.4: 54–7.

_______. 2008b. "Ecological Citizenship and a Plan for Sustainable Development." *City* 12.2: 237–44.

McDilda, Diane Gow. 2008. *365 Ways to Live Green: Your Everyday Guide to Saving the Planet.* New York: Adams Media.

McDonald, Darryl. 2005. Interview at the Okotoks Recycling Center, Okotoks, AB, 9 August.

McDonough + Partners. 1992. "The Hannover Principles: Design for Sustainability." Paper prepared for Expo 2000 – the World's Fair, Hannover, Germany.

_______. 2001. "Chicago City Hall Green Roof." http://www.mcdonoughpartners.com/projects/chicagoroof/default.asp?projID=chicago-roof (website discontinued – last accessed 12 June 2007).

McDonough Braungart Design Chemistry (MBDC). n.d. "Introduction to Cradle-to-Cradle Design Framework." http://www.chinauscenter.org/attachments/0000/0001/CradleDesign.pdf (accessed 17 June 2011).

McDonough, William. 2002a. "Dear Conscious Choice." http://www.mcdonough.com/writings/conscious_choice.htm (accessed 5 June 2004).

_______. 2002b. "The Living City: Nature, Design and the Greening of Chicago." http://www.mcdonough.com/writings/living_city.htm (accessed 5 June 2004).

_______. 2003. "A Field of Dreams: Green Roofs, Ecological Design and the Future of Urbanism." http://www.mcdonough.com/writings/field_of_dreams.htm (accessed 12 June 2009).

_______. 2004. "Principles, Practices and Sustainable Design: Toward a New Context for Building Codes." http://www.mcdonough.com/writings/principles_practices.htm (accessed 5 June 2009).

_______. 2007. "Huangbaiyu Cradle to Cradle Master Plan." http://web.archieve.org/20070304220322/http://www.mddonoughpartners.html (website discontinued – last accessed 12 June 2009).

McDonough, William and Michael Braungart. 1998. "The NEXT Industrial Revolution." TheAtlantic.com. http://www.theatlantic.com/doc/print/199810/environment (accessed 5 June 2009).

_______. 2002a. "Redefining Green." http://www.mcdonough.com/writings/redefining_green.htm (accessed 5 June 2004).

_______. 2002b. *Cradle to Cradle: Remaking the Way We Make Things.* New York: North Point Press.

_______. 2003. "Toward a Sustaining Architecture for the 21st Century: the Promise of Cradle-to-Cradle Design." *UNEP Industry and Environment*, April–September: 13–16.

McGray, Douglas. 2007. "Pop-Up Cities: China Builds a Bright Green Metropolis." *Wired Magazine* 15.5. http://www.wired.com/wired/archive/15.05/feat_popup.html (accessed 17 June 2011).

McKenzie, Funk. 2007. "China's Green Evolution." *Popular Science*, 2 July. http://www.popsci.com/scitech/article/2007-07/chinas-green-evolution (accessed 30 June 2010).

McKibben, Bill. 2010. *Earth: Making a Life on a Tough New Planet*. New York: Times Books/Henry Holt & Company.

Meadows, Dennis. 1972. *The Limits to Growth: A Report for the Club of Rome's Project on the Predicament of Mankind*. New York: Universe Books.

Miara, Jim. 2009. "Reviving Cities: Chattanooga, Camden, Cleveland, Providence, and Philadelphia are All in Various Stages of Being Rebranded." *Urban Land*, 25 November: 86–7.

Mitchell, William J. 1995. *City of Bits: Space, Place and the Infobahn*. Cambridge. MA: MIT Press.

Mokrycke, John. 1996. "New Life for Downtown Hamilton." *Hamilton Spectator*, 23 August: A13.

Molotch, Harvey L.. 1976. "The City as a Growth Machine: Toward a Political Economy of Place." *American Journal of Sociology* 82.2: 309–32.

Moore, Steven A. 2007. *Alternative Routes to the Sustainable City: Austin, Curitiba, and Frankfurt*. New York: Lexington Books.

Mortimer, Nigel D. and John F. Grant. 2008. "Evaluating the Prospects for Sustainable Energy Development in a Sample of Chinese Villages." *Journal of Environmental Management*. 87.2: 276–86.

Motavalli, Jim. n.d. "Chattanooga on a Roll: From America's Dirtiest City to its Greenest." *E: the Environmental Magazine*. http://www.emagazine.com/view/?426 (accessed 30 January 2010).

Mufson, Steve. 1985. "Third World Pleas on Commodity Prices Get No Sympathy in Developed Nations." *Wall Street Journal*, 2 October.

Mumford, Lewis. 1961. *The City in History: Its Origins, Its Transformations, and Its Prospects*. New York: Harcourt Brace Jovanovich.

Munasinghe, Mohan and Wilfrido Cruz. 1995. *Economywide Policies and the Environment: Lessons from Experience (World Bank Environment Paper No. 10)*. Washington, DC: The World Bank.

Nash, Roderick. 2001. *Wilderness and the American Mind*. New Haven, CT: Yale University Press.

Naughton, Barry. 2005. "The New Common Economic Program: China's Eleventh Five Year Plan and What It Means." *China Leadership Monitor* 16. http://www.hoover.org/publications/china-leadership-monitor/article/7332 (accessed 1 July 2011).

Naylor, Kenneth. 2003. "About Face: How Social Capital Transformed Chattanooga." *New American City* 3 (October): 38–41.

Nectoux, Francois and Yoichi Kurda. 1989. *Timber from the South Seas: An Analysis of Japan's Tropical Timber and its Environmental Impact*. Gland: World Wide Fund for Nature.

Neumeyer, Eric. 1999. *Weak versus Strong Sustainability: Exploring the Limits of Two Opposing Paradigms*. Northampton, MA: Edward Elgar.

NoCalgaryVeto website. http://www.nocalgaryveto.com (last accessed 17 June 2010).

NoCalgaryVeto. 2009. "Town Council Adopts a Contradictory Stance on Growth." http://nocalgaryveto.wordpress.com/2009/05/27/okotoks-town-council/ (accessed 23 February 2010).

Nordhaus, Ted and Michael Shellenberger. 2007. *Breakthrough: From the Death of Environmentalism to the Politics of Possibility*. New York: Houghton Mifflin.

Normile, Denise with Richard Stone. 2008. "China's Living Laboratory in Urbanization." *Science* 319: 740–43.

Odum, Eugene P. 1969. "The Strategy of Ecosystem Development." *Science* 164: 262–70.

______. 1989. *Ecology and Our Endangered Life-Support Systems*. Sunderland, MA: Sinauer Associates Inc.

Oikodrom SUCCESS website. http://www.oikodrom.org/?x=3.3 (last accessed 30 June 2010).

Ophuls, William. 1977. *Ecology and the Politics of Scarcity*. San Francisco: W. H. Freeman.

Orr, David W. 2002. *The Nature of Design: Ecology, Culture, and Human Intention*. New York: Oxford University Press.

Osborne, D. and T. A. Gaebler. 1992. *Reinventing Government: How the Entrepreneurial Spirit is Transforming the Public Sector*. New York: Perseus Press.

Paehlke, Robert C. 1989. *Environmentalism and the Future of Progressive Politics*. New Haven, CT: Yale University Press.

Parkinson, Sara and Mark Roseland. 2002. "Leaders of the Pack: An Analysis of the Canadian 'Sustainable Communities' 2000 Municipal Competition." *Local Environment* 7 (November): 411–29.

Parr, John. 1998. "Chattanooga: The Sustainable City." In Bruce Adams and John Parr (eds), *Boundary Crossers: Case Studies of How Ten of America's Metropolitan Regions Work*, 38–56. Baltimore, MD: Academy of Leadership Press.

Patterson, Don. 2010a. "Town's Water License Reaching its Capacity." *Western Wheel*, 6 January.

______. 2010b. "Future Growth in Jeopardy Without Water." *Western Wheel*, 13 January.

______. 2010c. "Town Needs to Re-evaluate Growth Plan, Says Expert." *Western Wheel*, 13 January.

Payer, Cheryl. 1974. *The Debt Trap: The International Monetary Fund and the Third World*. New York: Monthly Review Press.

Pierce, Neal and Curtis Johnson. 1998. "A Civic Vignette: The Chattanooga Story – From Troubled Raw River Town to Global Model." In Neal Pierce and Curtis Johnson (eds), *Boundary Crossers: Community Leadership in a Global Age*, 1–6. Baltimore, MD: Academy of Leadership Press.

Pearce, Will. 2005. Telephone interview, 1 December.

______. 2002. "'Sustainable Okotoks': Planning and Visioning Process." Municipal Excellence Network. http://www.menet.ab.ca/bins/view_practice.asp?pid=40 (website discontinued – last accessed 12 January 2005).

Pirages, Dennis. 1977. "Introduction: A Social Design for Sustainable Growth." In Dennis Pirages (ed.), *The Sustainable Society*, 1–13. New York: Praeger Publishers.

Pittsburgh Civic Design Coalition. 2006. "Report from Chattanooga." Civic Design Exchange, Pittsburgh Civic Design Coalition.

Poling, Jim. 1995. "48% Growth Forecast for New 25 Years." *Hamilton Spectator*, 30 May: B1.

Pollock, Rebecca M. and Graham S. Whitelaw. 2005 "Community-based Monitoring in Support of Local Sustainability." *Local Environment* 10 (June): 211–28.

Poovey, Bill. 2005. "Chattanooga: A Riverfront Transformed." *USA Today*, 16 May.

Porter, Charlene. 2000. "It Takes Us All, It Takes Forever." *Global Issues Electronic Journal*, March. http://usinfo.state.gov/journals/itgic/0300/ijge/gj-08.htm (website discontinued – last accessed 28 May 2007).

Portney, Kent E. 2003. *Taking Sustainable Cities Seriously: Economic Development, the Environment, and Quality of Life in American Cities*. Cambridge, MA: MIT Press.

Poster, Mark. 1995. "Cyberdemocracy: Internet and the Public Sphere." http://www.hnet.uci.edu/mposter/writings/democ.html (accessed 15 March 2003).

PowerPulse. 2003. "Advanced Vehicle Systems Closes its Doors." http://www.powerpulse.net/story.php?id=9378 (accessed 28 May 2010).

Pretty, J. 1995. "Participatory Learning for Sustainable Agriculture." *World Development* 23.8: 1247–63

Prigogine, Ilya and Isabelle Stengers. 1984. *Order Out of Chaos: Man's New Dialogue with Nature.* Boulder, CO: New Science Library.

Prokaska, Lee. 1995. "After the Revolution: The Rocky Road to the New Ontario." *Hamilton Spectator*, 4 October: B2.

Province of Alberta. 2008. "Land Use Framework." http://www.landuse.alberta.ca/documents/Final_Land_use_Framework.pdf (accessed 17 June 2010).

Public Citizen. 1991. "The Dolphin Case is Just One of Many: Examples of Other Recent Trade Disputes that Undermine Environmental, Health, and Safety Laws." Public Citizen, December.

Punter, John. 2004. *The Vancouver Achievement: Urban Planning and Design.* Seattle: University of Washington Press.

Putman, Yolanda. 2007. "CNE, Local Housing Developing Group Separate." *Chattanooga Times Free Press*, 20 September.

_______. 2009. "Stop Toxic Pollution Leader Leaves Group." *Chattanooga Times Free Press*, 21 June.

Putman, Yolanda and Jason Reynolds. 2009. "Homeownership Lags in Chattanooga." *Chattanooga Times Free Press*, 14 January.

Quail, Rick. 2009. "Sustainable Okotoks: 'Where to from Here.'" PowerPoint presentation. http://www.okotoks.ca/data/1/rec_docs/310_SOWTFHPres.pdf (accessed 26 February 2010).

Quinn, Michael. n.d. "Where Land Meets Water: Riparian Challenges for the New Millennium – Health and the Urban Riparian Zone: The Sustainable Okotoks Project." http://www.tucanada.org/riparian99/quinn.htm (accessed 4 June 2004).

Ranulf, Svend. 1964. *Moral Indignation and Middle Class Psychology.* New York: Schocken Books.

Rawls, John. 1989. *A Theory of Justice.* Cambridge, MA: Harvard University Press.

Redclift, Michael. 1984. *Development and the Environmental Crisis: Red or Green.* London: Methuen.

_______. 1987. *Sustainable Development: Exploring the Contradictions.* New York: Routledge.

Regional Municipality of Hamilton-Wentworth. 1996a. "Signposts on the Trail to VISION 2020: Hamilton-Wentworth's Sustainability Indicators." October.

_______. 1996b. "Sustainable Community Decision-making Guide." Paper by the Administrative Services Committee, 23 May.

_______. 1996c. "VISION 2020 Evaluation (PLA 96-036)." Paper by the Economic Development and Planning Committee, 9 September.

Register, Richard. 2002. *Ecocities: Building Cities in Balance with Nature.* Berkeley, CA: Berkeley Hills Books.

Reich, R. 1991. *The Work of Nations.* New York: Vintage Books.

Reid, Herbert G. and Betsy Taylor. 2010. *Recovering the Commons: Democracy, Place, and Global Justice.* Chicago: University of Illinois Press.

_______. 2010. *Recovering the Commons: Democracy, Place, and Global Justice.* Champaign, IL: University of Illinois Press.

Reynolds, Jason. 2007. "CNE Sells 2 Apartment Units for $2.4 Million." *Chattanooga Times Free Press*, 7 November.

_______. 2008. "Chattanooga Neighborhood Enterprise Inc. Seeks Director to Help Refocus Mission." *Chattanooga Times Free Press*, 16 May.

Reynolds, Jason and Dave Flessner. 2007. "CNE Continues to Narrow Mission with Less Funding." *Chattanooga Times Free Press*, 30 May.

Rheingold, H. 1993. *The Virtual Community: Homesteading on the Electronic Frontier.* Reading, MA: Addison-Wesley.

Rich, Bruce. 1994. *Mortgaging the Earth: The World Bank, Environmental Impoverishment, and the Crisis of Development.* Boston: Beacon Press.

RMI Hypercar Center website. http://www.hypercarcenter.org (last accessed 24 November 2003).

Roberts, Dexter. 2008. "China Prepares for Urban Revolution." *Bloomberg Businessweek*, 13 November. http://www.businessweek.com/globalbiz/content/nov2008/gb20081113_305364.htm?chan=rss_topEmailedStories_ssi_5 (accessed 24 November 2003).

Robertson-Rehberg, Eileen. 2008. *Housing: 2008 State of Chattanooga Region Report.* Chattanooga, TN: Ochs Center for Metropolitan Studies.

Rogge, Mary E. 1998. "Toxic Risk, Community Resilience, and Social Justice in Chattanooga, Tennessee." In Marie D. Hoff (ed.), *Sustainable Community Development: Studies in Economic, Environmental, and Cultural Revitalization*, 105–22. New York: Lewis Publishers.

Rogge, Mary E., Kimberley Davis, Deborah Maddow and Milton Jackson. 2005. "Leveraging Environmental, Social, and Economic Justice at Chattanooga Creek: A Case Study." *Journal of Community Practice* 13.3: 33–53.

Roseland, Mark. 1998. *Toward Sustainable Communities: Resources for Citizens and Their Governments.* Gabriola Island, BC: New Society Publishers.

_______. 2005. *Toward Sustainable Communities: Resources for Citizens and Their Governments.* Gabriola Island, BC: New Society Publishers.

Ruiz, Richard. 2006. "Visions of Ecopolis." *Economist* 380.8496: 20–3.

Sachs, Wolfgang. 1992. "Global Uniformity." *Resurgence*, March/April.

_______. 1993. *Global Ecology: A New Arena of Political Conflict.* Atlantic Heights, NJ: Zed Books.

Schattschneider, E. E. 1960. *The Semi-Sovereign People: A Realist's View of Democracy in America.* New York: Holt, Rinehart and Winston.

Schifferes, Steve. 2007. "Chinas Eco-City Faces Growth Challenge." BBC News, 5 July. http://chinadigitaltimes.net/2007/07/chinas-eco-city-faces-growth-challenge-steve-schifferes/ (accessed 17 June 2010).

Schnaiberg, Allan. 1980. *The Environment: From Surplus to Scarcity.* New York: Oxford University Press.

Schwartz, N. 1985. "Communitarian Citizenship: Marx &Weber on the City." *Polity* 17 (Spring): 530–48.

Shafer, Sarah. 2007. "Trouble for China's Model Green City." *China Environmental News Digest*, 11 May. http://china-environmental-news.blogspot.com/2007/05/trouble-for-china-model-green-city.html (accessed 5 June 2007).

Shafer, Sarah and Anne Underwood. 2005. "Building in Green." *Newsweek International*, 26 September–3 October.

Shaw, Jane S. and Ronald D. Utt. 2001. *A Guide to Smart Growth: Shattering Myths, Providing Solutions.* Washington, DC: Heritage Foundation.

Shaw, V. L., A. J. Hunter and N. D. Mortimer. 2007. "Solar World Congress 2007: Sustainable Energy Development for Rural China." Paper presented at the International Solar Energy Congress, Beijing, September 18–21.

Shepard, Paul. 1982. *Nature and Madness.* San Francisco: Sierra Club Books.

Shipley, Anna Monis and R. Neal Elliott. 2006. *Ripe for the Picking: Have We Exhausted the Low-hanging Fruit in the Industrial Sector?* Washington, DC: American Council for an Energy-Efficient Economy.

Shiva, Vandana. 1993. "The Greening of the Global Reach." In Wolfgang Sachs (ed.), *Global Ecology: A New Arena of Political Conflict*, 149–56. Atlantic Heights, NJ: Zed Books.

Simon, Julian. 1981. *The Ultimate Resource*. Princeton, NJ: Princeton University Press.

Sohn, Pam. 2008. "Chattanooga: Metropolitan Region Leaving a Larger Carbon Footprint." *Chattanooga Times Free Press*, 28 October.

_______. 2010. "New Look with a Green Tint." *Chattanooga Times Free Press*, 1 January.

Sorel, Georges. 1961. *Reflections on Violence*. New York: Collier-Macmillan.

Statistics Canada. 1991. *Census of Canada (1991)*. Available online at http://www.statcan.gc.ca (last accessed 1 August 2011).

_______. 2007. "Community Profiles: Okotoks – Age Characteristics for Both Sexes (2006)." http://www12.statcan.ca/census-recensement/2006/dp-pd/prof/sip/Rpeng.cfm?TABID=1&LANG=E&APATH=3&DETAIL=0&DIM=0&FL=A&FREE=1&GC=0&GK=0&GRP=1&PID=97614&PRID=0&PTYPE=97154&S=0&SHOWALL=0&SUB=0&Temporal=2006&THEME=80&VID=0&VNAMEE=&VNAMEF= (accessed 17 June 2011).

Stevens, Dannelle D. and Antonia J. Levi. 2004. *Introduction To Rubrics: An Assessment Tool To Save Grading Time, Convey Effective Feedback and Promote Student Learning*. Herndon, VA: Stylus Publishing.

Stiglitz, Joseph E. 2010. *Freefall: America, Free Markets, and the Sinking of the World Economy*. New York: W. W. Norton & Company.

Stockwell, J. R. and J. W. Sorenson. 1994. *The Chattanooga TNGA Metropolitan Statistical Area (MSA) Toxic Release Inventory – Geographic Information System (TRI-GIS) Comparative Risk Screening Analysis*. Atlanta, GA: US Environmental Protection Agency.

Stone, Clarence N. 1989. *Regime Politics: Governing Atlanta, 1946–1988*. Lawrence, KS: University of Kansas Press.

_______. 1993. "Urban regimes and the Capacity to Govern: A Political Economy Approach." *Journal of Urban Affairs* 15: 1–28.

_______. 2005. "Looking Back to Look Forward: Reflections on Urban Regime Analysis." *Urban Affairs Review* 40 (January): 309–41.

Stone, Richard. 2006. "Villagers Drafted into China's Model of 'Sustainability.'" *Science* 312: 36.

Streeter, April. 2006a. "A Tale of Two Cities: China's sprawling Cityscapes are Starting to Sprout Some Eco-Alternatives." *GreenFutures: The Sustainable Solutions Magazine*, 2 November.

_______. 2006b. "Big Trouble in Rural China." Sustainable Industries, 8 April. http://www.sustainableindustries.com/sijnews/2713166.html (accessed 30 June 2010).

Sudjic, Dejan. 2006. "Making Cities Work: China." BBC News, 21 June. http://news.bbc.co.uk/1/hi/world/asia-pacific/5084852.stm (accessed 9 December 2009).

Surowiecki, James. 2002. "Waste Away." *New Yorker*, 6 May. http://www.newyorker.com/archive/2002/05/06/020506ta_talk_surowiecki (accessed 30 June 2010).

Sustainable Okotoks. 2010. "Water Management Plan." http://www.town.okotoks.ab.ca/suswater.html (website discontinued – last accessed 17 June 2010).

_______. 2005. "Public Participation Process: Community Survey 1997, 1998, 2000, 2003." http://www.town.okotoks.ab.ca/sussurvey.html (website discontinued – last accessed 12 December 2004).

Task Force on Sustainable Development, Regional Municipality of Hamilton-Wentworth. 1993. "Implementing Vision 2020: Detailed Strategies and Actions Creating a Sustainable Region." Regional Municipality of Hamilton-Wentworth, January.

Taylor, Charles. 1971. "Interpretation and the Sciences of Man." *Review of Metaphysics* 25 (September): 3–51.

Tharp, William. 2008. *Environment – 2008 State of Chattanooga Report.* Chattanooga, TN: Ochs Center for Metropolitan Studies.

Tools of Change. 2000. "Case Studies: Okotoks" http://www.toolsofchange.com/English/CaseStudies/?ID=149 (accessed 17 June 2010).

Town of Okotoks. 2005. "Let's Talk Trash!...paper, cans, plastics and other recyclables!" Fact Sheet Number 2.

Town of Okotoks. n.d. "Okotoks Integrated Wastewater Treatment System." Town of Okotoks Planning Services.

_______. 1998. "Municipal Development Plan: The Legacy Plan." Town of Okotoks Planning Services, September.

_______. 2004. "1998–2003: Municipal Development Plan Review." Town of Okotoks Planning Services, March.

_______. 2009a. "Sustainable Okotoks: 'Where to from Here?'" http://www.okotoks.ca/default.aspx?cid=748&lang=1 (accessed 26 February 2010).

_______. 2009b. "Sustainable Okotoks: 'Where to from Here?'" [Red, Green, and Yellow Scenarios]. http://www.okotoks.ca/data/1/rec_docs/310_SOWTFHPres.pdf (accessed 17 December 2009).

_______. 2009c. "Community Household Survey Report." http://www.okotoks.ca/data/1/rec_docs/448_G.1_Attach_1_Household_Survey_Results.pdf (accessed 26 February 2010).

_______. 2009d. "Town of Okotoks to Maintain Population Cap of 30,000." Press release, 26 May.

_______. 2010. "Integrated Sustainability Framework." http://www.okotoks.ca/data/1/rec_docs/925_Okotoks_ICSP_Framework_March_20_2010.pdf (accessed 17 June 2010).

Town of Okotoks website http://www.okotoks.ca/home.aspx (last accessed 17 June 2011).

Toy, Mary-Anne. 2006a. "China's First Eco-village Proves a Hard Sell." *The Age*, 26 August. http://www.theage.com.au/news/world/chinas-first-ecovillage-proves-a-hard-sell/2006/08/25/1156012740582.html?page=fullpage (accessed 30 June 2010).

_______. 2006b. "Green Dream Vanishes in Puff of Reality." *Sydney Morning Herald.* http://www.smh.com.au/news/world/green-dream-vanishes-in-puff-of-reality/2006/08/25/1156012739220.html (accessed 30 June 2010).

Uliano, Sophie. 2008. *Gorgeously Green: 8 Simple Steps to an Earth-friendly Life.* New York: Harper Paperbacks.

United Nations. 1992. "Agenda 21." http://www.un.org/esa/dsd/agenda21/res_agenda21_00.shtml (accessed 17 June 2011).

United Nations Human Settlements Programme (UN-HABITAT). 2006. *State of the World's Cities 2006/2007: 30 Years of Shaping the Habitat Agenda.* London: Earthscan.

United Nations World Commission on Environment and Development (Brundtland Commission). 1987. *Our Common Future.* New York: Oxford University Press.

Valkenier, Elizabeth. 1982. *The Soviet Union and the Third World.* New York: Praeger Press.

Vance, James E. Jr. 1990. "The Face of the Medieval City." *The Continuing City: Urban Morphology in Western Civilization.* Baltimore, MD: Johns Hopkins University Press. 111–71.

Volkswagen, Das Werk. 2009. "Green from the Start – Volkswagen's Commitment to Sustainable Future in Chattanooga." Press folder. http://www.volkswagengroupamerica.com/media/docs/20090514_sustainability.pdf (accessed 28 May 2010).

Wackernagel, Mathis and William Rees. 1998. *Our Ecological Footprint: Reducing Human Impact on the Earth.* Gabriola Island, BC: New Society Publishers.

Wang, Yongqiang. 2006. "The Renewable Village – A Case Study in Sustainable Development." *China Pictorial*, September. http://www.rmhb.com.cn/chpic/htdocs/english/200609/5-2.html (accessed 30 June 2010).

Weber, Max. 1958. *The City*. Trans. and ed. Don Martindale and Gertrud Neuwirth. Glencoe, IL: Free Press.

Wexler, Kathryn. 2001. "City Wants Electric Buses by Nov. 30." *St. Petersburg Times*, 3 November.

Wilkins, Charles. 1993. "Steeltown Charts a New Course." *Canadian Geographic* 113 (July/August): 42–55.

Williams, Dan. 2006. "China as Green Lab." *Harvard Business Review*, February: 38–40.

Williams, Raymond. 1975. *The Country and the City*. New York: Oxford University Press.

Wilson, Matt. 2007. "Chattanooga Creek Cleanup Effort Coming to a Close." *Chattanooga Times Free Press*, 25 August.

World Bank. 1992. *World Development Report 1992, Development and the Environment*. New York: Oxford University Press.

Yanarella, Ernest J. 1998. "Plato Meets Lawnmower Man in the Virtual Polis: The Case of PS 776." *P.S.: Political Science and Politics* 31: 792–6.

_______. 1999. "Local Sustainability Programs in Comparative Perspective: Canada and the USA." *Local Environment* 4 (June): 209–23.

_______. 2001. *The Cross, the Plow and the Skyline: Contemporary Science Fiction and the Ecological Imagination*. Boca Raton, FL: Brown Walker Press.

_______. 2009. "Ecological Wisdom and Postmodern Defiance in Italo Calvino's Invisible Cities." *Wemarpolis* 1.2: 65–80. http://www.weimarpolis.net/docs/volume1/issue2/Yan_Full_Article.pdf (accessed 22 February 2010).

Yanarella, Ernest J. and Robert W. Lancaster. n.d. *Getting from Here to There: The Politics of Urban Sustainability in North America*. Forthcoming.

Yanarella, Ernest J. and Richard S. Levine. 1992a. "Five Operating Principles for Sustainable Cities." A paper prepared for the American Solar Energy Society Conference, Cocoa Beach, Florida, May.

_______. 1992b. "Does Sustainable Development Lead to Sustainability?" *Futures*. 24 (October): 759–74.

———. 1992c. "The Sustainable Cities Manifesto: Pre-Text, Text, and Post-Text." *Built Environment* 18 (December): 301–13.

_______. 2000. "The Space of Flows, the Rules of Play, and Sustainable Urban Design: the Sustainability Game as a Tool of Critical Pedagogy in Higher Education." *International Journal of Sustainability in Higher Education* 41: 48–66.

_______. 2008. "Don't Pick the Low-Hanging Fruit? Counterintuitive Policy Advice for Sustainability." *Sustainability: The Journal of Record* 1.4: 256–60.

Yanarella, Ernest J., Richard S. Levine and Robert W. Lancaster. 2009. "Green" vs. Sustainability: From Semantics to Enlightenment. *Sustainability: The Journal of Record*. 2.5: 296–302.

Yoko, Greg. 2007. "Partnering is Core Philosophy in Chattanooga." *Sustainable Urban Redevelopment* 3.1. http://www.surmag.com/index.php?option=com_content&task=view&id=4&Itemid=7&ed=2 (accessed 28 May 2010).

Youtie, Jan. 1999. "Case Study: Chattanooga – Decentralized Approach to Sustainable Development" In Jan Youtie (ed.), *Innovative Local Economic Development Programs: A Report*, 37–44. Washington, DC: Economic Development Administration, US Department of Commerce.

Yudelson Jerry. 2007. *Green Building A-Z: Understanding the Language of Green Building*. Gabriola Island, BC: New Society Publishers.

Zarsky, Lyuba. 1991. *International Trade and Ecologically Sustainable Development: Policy Directions*. Melbourne: Australian Conservation/World Wide Fund for Nature.

Ziegler, Edward H. 2006. "China's Cities, Globalization, and Sustainable Development: Comparative Thoughts on Urban Planning, Energy, and Environmental Policy." *Washington University Global Studies Law Review* 5.2: 295–322.

Zovanyi, G. 2001. *Growth Management for a Sustainable Future: Ecological Sustainability as the New Growth Management Focus of the 21st Century*. New York: Praeger.

INDEX

D

E

F

Q

R

T

CPSIA information can be obtained at www.ICGtesting.com
Printed in the USA
BVOW062255260212

283782BV00002B/2/P